Community Pest Management in Practice

Tanya M. Howard · Theodore R. Alter
Paloma Z. Frumento · Lyndal J. Thompson

Community Pest Management in Practice

A Narrative Approach

 Springer

Tanya M. Howard
Australian Centre for Agriculture and Law
University of New England
Armidale, NSW, Australia

Theodore R. Alter
Center for Economic and Community
  Development
Pennsylvania State University
University Park, PA, USA

Paloma Z. Frumento
Center for Economic and Community
  Development
Pennsylvania State University
University Park, PA, USA

Lyndal J. Thompson
Department of Education and Training
Canberra, ACT, Australia

ISBN 978-981-13-2741-4   ISBN 978-981-13-2742-1   (eBook)
https://doi.org/10.1007/978-981-13-2742-1

Library of Congress Control Number: 2018955921

This Springer imprint is published by the registered company Springer Nature Singapore Pte Ltd.
The registered company address is: 152 Beach Road, #21-01/04 Gateway East, Singapore 189721,
Singapore

In memory of Harley West.

Foreword

The first human colonisers of Australia contributed to one of the great ages of mega-fauna extinction. In the twenty-first century, a similar impact is being made by a wide range of invasive animal and weed species, evidenced by obvious and rapid landscape-scale ecological and socio-political change. This important book tells stories of people and community engagement in the face of this rapid change; and their efforts to manage, if not eradicate, the impact of invasive species on behalf of the wider community.

This book is made up of three case studies and twelve personal narratives of vision and energy, passion and commitment, tragedy and loss, the mundane and the routine, frustration and joy, learning and leading, and the romance and mystique of living and working in unique landscapes across a continent tamed yet somehow remaining ecologically, institutionally and socially wild at heart. Essentially, this book tells stories of peoples' sense of place and self, sometimes revealed explicitly and sometimes more implicitly in these very personal stories.

While it is true that a remarkable group of social scientists has enabled these stories to come forth, it is the voices of the subjects—front-line practitioners, farmers, community members and citizens—that stand out. I have worked with such people for decades and stand in awe of their contribution to the Australian environment and sustainable agriculture. But by the time I reached the end of this book, it was clear that the authors themselves had made significant transitions and that they too are subjects, and not just tellers, of an amazing story.

So knowing the reader will come away learning much from the practitioners stories, and hopefully also being as inspired as I was, this Foreword focuses on the story of the authors and the voice of the social sciences they most effectively channel. And my word, this book carries significant weight in bringing to life the voice of the social sciences and the unique role they play in making sense of the complex relationship between policy, science, experience, world views and practice.

Perhaps surprisingly, there is no single, authoritative definition of *social science*. Like the term *biophysical science*, social science represents a collection of disciplines, in this case disciplines that deal with society and the human condition. The key element is a focus on human behaviour and those disciplines that explore human behaviour in some detail. A secondary element is a focus on those institutions which humans create and the culture related to the various ways that people live. I can think of few better examples of a text that represents the many dimensions of human behaviour shaping and shaped by the institutions of society.

With some humility, the opening chapter of the book calls encourages the reader to make what we will of the text by reflecting upon the lived experience of practitioners and case study communities. The book makes no explicit attempt to spell out authoritative lessons but suggests that like all truly compelling lessons in life, you have to experience them for yourself to come to your own understanding and definition. This is where this book comes in handy, offering us a glimpse into the lives of people as they reflect on and share their life lessons, and in doing so also allowing the reader to come to a more personal understanding of biosecurity in action.

If I had written this Foreword 20 years ago, I would have added to that preceding sentence '*in the context of environmental and natural resource management.*' These days, the window looking out over our landscapes is less that of an *environmental* pane than a *biosecurity* one, and with this subtle shift in policy discourse in Australia comes a need to re-evaluate the contribution of the social sciences from both academic and utilitarian perspectives. Social sciences provide hope for dealing with complex and intractable problems where the interests of nature, communities and economies collide and often compete. Indeed, many of its disciplines have evolved to become relevant to environmental, sustainable agriculture and natural resource management issues, making them highly relevant to addressing the issues of biosecurity.

In related scribblings published elsewhere, the authors have stated: "Although the language of *shared responsibility* is being implemented in different jurisdictions as a regulated citizen biosecurity obligation, it is poorly understood by community members, and this has significant implications for effective implementation" (Howard et al. 2018, p. 1). Institutional advocacy of shared responsibility is an acknowledgement that non-government stakeholders are crucial to biosecurity solutions. However, this acknowledgement does not always recognise that stakeholders may see the world more broadly than through the science-centric lens often applied to biosecurity management. Involving the stakeholders of natural resource management agencies in decisions and/or management is an activity that has grown considerably over the past two decades. Often badged as *community engagement*, the term is used loosely and rhetorically by many public institutions, and as a result questions about who should participate and how they ought to participate are not always dealt with effectively. An important job of social and behavioural scientists in natural resource management, and now too in the biosecurity field, is to assist citizens, government officials, and industry leaders to co-create the enabling settings and participatory processes required for effective public discourse. Public spaces are needed that allow the expertise, experience, and voices of multiple parties to be heard and synthesised in naming and framing the issues at hand, as well as figuring how best to address these issues going forward at a particular point in time in particular situations. This work is ongoing and requires constant attention and nurturing by all parties. And here is where the authors of this book have not merely observed, interviewed and relayed practitioner stories and case studies. Through their own reflections, having shared the lived experience of practitioners and community members over a number of years, the researchers have become essential partners in the ongoing evolution of the human dimensions of Australia's biosecurity challenge.

This book certainly demonstrates their credentials, in a finely crafted tale of tales.

Canberra, Australia   Associate Professor Richard Price, Ph.D.
                                                                                              FAIST, GAIDC

Reference

Howard, T. M., Thompson, L. J., Frumento, P. Z., & Alter, T. (2018). Wild dog management in Australia: An interactional approach to case studies of community-led action. *Human Dimensions of Wildlife, 23*(3). https://doi.org/10.1080/10871209.2017.1414337.

Preface

Across Australia and around the world, governments, industry bodies and community members are struggling to tackle the serious and complex impacts of established pest species. Pest species take hold in landscapes that are segmented by land-tenure boundaries, where there are fragmented governance regimes and short-term funding and planning cycles. Current management and control approaches are informed by technical expertise in species ecology. Successful implementation, however, also requires sustained and coordinated community action. This is particularly true for agricultural communities, where the impact of pest species can have a disastrous impact on livelihoods and take a heavy toll on individual mental health and community cohesiveness.

Challenges to effective community action for coordinated pest management in Australia include:

- A vast landmass with small populations outside urban areas allowing pests to spread and adapt across the country;
- Changing work patterns and demographic drift to urban areas leading to reduced volunteer capacity for pest control;
- Many weed and pest species have become widely established and cannot be eradicated, creating a persistent, recurring management issue;
- An over-reliance on technical expertise as the solution to pest problems leading to neglect of citizen knowledge in defining the problem and finding workable solutions.

Other barriers include:

- the cost of ongoing pest control and who bears this;
- the challenge of coordinating control across different land tenures;
- the lack of long term planning and funding by government agencies; and
- the power imbalances that occur in any community when money and politics is involved.

Until recently, the study of pest management has largely been conducted as a branch of wildlife ecology or agricultural science. Scientific methods have been employed to study pest species, their habitat, ecological factors and the species impact on biodiversity, agricultural productivity and landscape amenity. There have been waves of investment in control technology and on-ground implementation; however, the human dimension of these problems, such as the social dynamics that underpin community action, has been only recently been identified as a necessary avenue for research in the pest management context (Ballard 2006; Miller 2009). This growth in human dimensions research stems from a growing awareness of the role individual and collective human behaviour plays in the success or failure of pest management (Manfredo et al. 2009; Dickman et al. 2013).

The challenge of achieving landscape-scale coordinated action for pest management has also stimulated social research into the policy and practice settings that can support voluntary action by private citizens (Marshall et al. 2016; Martin 2016). This research shows that policy interventions are often based on assumptions of both scientific and social values regarding pest species (Dickman 2010). However, when faced with difficult decisions in complex contexts, better decisions and more effective action result from combining specialised knowledge with community values and local knowledge (Sjölander-Lindqvist et al. 2015; Almeida-LeÑEro et al. 2017). This has led to calls for increased community engagement as policy-makers and practitioners realise that incorporating community values into decisions about pest management can increase the likelihood of community acceptance and action (Coralan 2006; Fischer 2005).

This book emerged from a broader research program focused on deepening awareness of the human dimensions of invasive species control. Funded by the Australian Invasive Animals Cooperative Research Centre from 2013 to 2017, the program included research into human behaviours; the design and implementation of management regimes; and the way that community members participate (or not) in these regimes. Within this broader program, our research team developed a specific focus on community building and collective-action strategies for pest management. Drawing on theories of democratic practice, social change, community development and political economy, we began to frame established pest management as a *problem of community*. "Community" is broadly defined to include farmers, land managers and the stakeholders they engage with for pest species management. While pest species can be a trigger for community formation, particularly when there is a collective recognition and acceptance of a threat (Everts 2015), the persistent nature of pest species and the ecological characteristics that encourage their dispersal and survival in the landscape also require widespread and ongoing community action.

We argue that moving beyond community formation to community action requires development of a *shared vision* of both the threat and possible solutions. Once community members are able to develop a shared vision, they are better equipped to make a *shared commitment* to take collective action for pest species management. Sharing stories of pest management is a uniquely human way to build this collective vision and commitment.

For us, attempts to facilitate community-led action for pest species management are strengthened by theories and practices of community building. These include acceptance of diverse values and the capacity to work across difference; awareness of power differentials and active management of these dynamics in any community engagement efforts; recognition of the tension between expertise and local knowledge in understanding and responding to pest threats; and a conviction that social and environmental change begins with individual reflection and learning. In this book, we place theory and practice in close proximity, considering what each can tell us about the process of community formation and collective action for pest management. Through a collection of 12 unique and engaging practitioner stories and three case studies of wild dog management groups, we present a narrative enquiry into contemporary pest management in Australia. The practice stories provide an in-depth look into the motivation and thinking of individuals involved in pest management and community engagement. The wild dog management group case studies investigate how community groups engage in the specific problem of wild dog management and the factors that shape their collective response. These practice stories and case studies explore how individuals and communities perceive pest species and how they frame the "problem" of management.

In combination, these narratives provide a novel perspective on how interaction between practitioners, private land managers and other stakeholders might impact on the individual and collective framing of pest species as a problem and the possible consequences for management strategies. We chose a narrative enquiry because it allowed us to explore individual and collective human experience of pest management (Webster and Mertova 2007). Narrative enquiry is not concerned with seeking causal explanations for social phenomena; it is concerned with under-standing social experience in order to develop a platform for the shared under-standing that we believe is particularly relevant to questions of community action (Bruner 1991; Ollerenshaw and Creswell 2002). Throughout our analysis, we consider what these stories might tell us about the human dimensions of Australian landscape management. We find that the narratives and case studies are animated by a range of social, economic and political factors which may stimulate, sustain or stymie collective community action for pest management.

What Follows

The book is organised into four parts.

- Chapters 1 and 2 establish the foundations for the research project: Chap. 1 explores the broader social and cultural context of the investigation; Chap. 2 describes the narrative enquiry methodology in detail.
- *Part I (Practitioner Profiles)* presents 12 individual Practitioner Profiles and an accompanying meta-analysis.
- *Part II (Wild Dog Groups)* includes three unique case study narratives, each accompanied by a narrative analysis. These include: Chap. 17 (Introduction to the cases), Chap. 18 (Mount Mee, Queensland); Chap. 19 (Ensay and Swifts Creek, Victoria); and Chap. 20 (Northern Mallee, Western Australia). Chapter 21 presents an integrative analysis of all three cases.
- Final conclusions are made in *Part III*, Chap. 22.

A note on reading the narratives: In shaping this collection of narratives, we have resisted the urge to present an authoritative account of "community engagement" in the form of best practice guidelines or instructions. We have embraced a variety of experiences, recognising the pluralistic world views that underpin individual practice and shape group experience. This collection brings together the voices of community members and leaders in the field, and provides a valuable repository of their experiences. Their stories can be regarded as learning resources and are presented in full so readers can contemplate and draw lessons as they see fit. The narratives are accompanied by analysis in which the research team presents our interpretation of these stories and discusses what we regard as significant insights based on this interpretation. These insights are grounded in our world view and theoretical framework which sees knowledge as constructed through debate and dialogue across different viewpoints.

The practitioner profiles were developed following the method outlined in the Peters, Alter and Schwartzbach text *Democracy and Higher Education*. Individual interviews were conducted with 12 practitioners involved in working with community members to manage a range of pest species. Interviews were guided by five open-ended questions that drew out their personal stories and positioned their professional experience within this broader context. The resulting transcripts were then edited, reshaped and re-storied in consultation with the practitioner, to develop a first-person narrative that presented a cohesive story of their practice and experience over the course of their career (Ollerenshaw and Creswell 2002). In these stories: "practitioners speak of what they did and learned in particular moments of their work and lives. They provide us with windows into the situational and contextual work and experiences of people who are engaged in civic life" (Peters et al. 2010).

The wild dog management group case studies combined documentary sources with in-depth interview data to develop a narrative account of each wild dog management group at a particular moment in time. The research aimed to explore the lived experience of those engaged in wild dog management, generating rich qualitative data grounded in the specific circumstances and context of the subject (Flyvbjerg 2001). Because group experience is co-created through individuals' interactions with each other, the use of narrative enquiry allowed us to access the group's shared understanding of their collective effort to manage wild dog threats (Black 2008).

Throughout this book, readers will be exposed to new and exciting ways of talking and thinking about community engagement, collective action and pest management. They will also encounter personal stories and professional philosophies of community engagement, generously shared by individuals working at the front line of community pest management. These accounts draw attention to the complex social dynamics of community formation that underpins the progress from threat awareness to collective action. In this way, our work contributes to the identified need to understand in more detail how community is produced, or not, in the context of pest species (Everts 2015). Through a combination of theoretical analysis and practical wisdom, we begin by emphasising that:

> The work of [community] engagement is happening already. The challenge is less about getting people to embrace a new idea and more about accentuating, promoting, developing, and celebrating the work that is already underway. (Fear et al. 2006, p. 18)

Our hope is that engagement practitioners, policy officers, pest managers and community members will be stimulated by these narratives to reflect on their own experience and further develop their personal and professional understanding of community action. We invite our readers to join us in the ongoing development of critical engagement theory and practice for community pest management.

Armidale, Australia   Tanya M. Howard
University Park, USA   Theodore R. Alter
University Park, USA   Paloma Z. Frumento
Canberra, Australia   Lyndal J. Thompson

References

Almeida-LeÑEro, L., Revollo-FernÁNdez, D., Caro-Borrero, A., Ruiz-MallÉN, I., Corbera, E., Mazari-Hiriart, M. & Figueroa F. (2017). Not the same for everyone: Community views of Mexico's payment for environmental services programmes. *Environmental Conservation, 44* (3), 201–211.

Ballard, G. (Ed.). (2006). Social drivers of invasive animal control. In *Proceedings of the Invasive Animals CRC workshop on social drivers of invasive animal control, 26th–27th July 2006*. Adelaide, SA: Invasive Animals Cooperative Research Centre, Canberra.

Black, L. W. (2008). Deliberation, storytelling, and dialogic moments. *Communication Theory, 18* (1), 93–116. https://doi.org/10.1111/j.1468-2885.2007.00315.x.

Bruner, J. (1991). The narrative construction of reality. *Critical Inquiry, 18*(1), 1–21.

Coralan, M. S. (2006). Science, expertise and the democratization of the decision-making process. *Society & Natural Resources, 19*, 661–668.

Dickman, A. J. (2010). Complexities of conflict: The importance of considering social factors for effectively resolving human–wildlife conflict. *Animal Conservation, 13*(5), 458–466. https://doi.org/10.1111/j.1469-1795.2010.00368.x.

Dickman, A., Marchini, S., & Manfredo, M. (2013). The human dimension in addressing conflict with large carnivores. In D. W. Macdonald & K. J. Willis (Eds.), *Key topics in conservation biology 2* (pp. 110–126). Oxford, UK: Wiley.

Everts, J. (2015). Invasive life, communities of practice, and communities of fate. *Geografiska Annaler: Series B, Human Geography, 97*(2), 195–208.

Fear, F. A., Rosaen, C. L., Bawden, R. J. & Foster-Fishman, P. G. (2006). *Coming to critical engagement: An authoethnographic exploration*. Maryland, USA: University Press of America.

Fischer, F. (2005). *Citizens, experts and the environment: The politics of local knowledge* (1st ed.). Durham, NC: Duke University Press.

Flyvbjerg, B. (2001). *Making social science matter: Why social inquiry fails and how it can succeed again*. Cambridge, UK: Cambridge University Press.

Manfredo, M. J., Vaske, J. J., Brown, P. J., Decker, D. J., & Duke, E. A. (2009). *Wildlife and society: The science of human dimensions*. Washington, DC: Island Press.

Marshall, G. R., Coleman, M. J., Sindel, B. M., Reeve, I. J., & Berney, P. J. (2016). Collective action in invasive species control, and prospects for community-based governance: The case of serrated tussock (*Nassella trichotoma*) in New South Wales, Australia. *Land Use Policy, 56*, 100–111. https://doi.org/10.1016/j.landusepol.2016.04.028.

Martin, P. (2016). Ecological restoration of rural landscapes: Stewardship, governance, and fairness. *Restoration Ecology, 24*(5), 680–685. https://doi.org/10.1111/rec.12411.

Miller, K. K. (2009). Human dimensions of wildlife population management in Australasia—History, approaches and directions. *Wildlife Research, 36*, 48–59.

Ollerenshaw, J. A., & Creswell, J. W. (2002). Narrative research: A comparison of two restorying data analysis approaches. *Qualitative Inquiry, 8*(3), 329–347.

Peters, S., Alter, T. R., & Schwartzbach, N. (2010). *Democracy and higher education: Traditions and stories of civic engagement*. East Lansing, MI: Michigan State University Press.

Sjölander-Lindqvist, A., Johansson, M., & Sandström, C. (2015). Individual and collective responses to large carnivore management: The roles of trust, representation, knowledge spheres, communication and leadership. *Wildlife Biology, 21*(3), 175–185. https://doi.org/10.2981/wlb.00065.

Webster, L., & Mertova, P. (2007). *Using narrative inquiry as a research method*. London, UK: Routledge.

Acknowledgements

We gratefully acknowledge the support of the Invasive Animals Cooperative Research Centre (IA-CRC) which funded this research as part of the multi-disciplinary research program "Facilitating collective action". This program was led by the Australian Centre for Agriculture and Law at the University of New England, in partnership with the Pennsylvania State University (USA) and Griffith University (Australia).

We thank all of the research team for their friendship and intellectual support over the course of the research: in particular, Program Leader Prof. Paul Martin and theme leaders Prof. Don Hine and Prof. Daryl Low Choy. Thank you to post-doctoral research fellow Dr. Patty Please for her wise counsel and creative interest in our work. The team also benefitted from the practical contributions and intellectual curiousity of Ph.D. students Katrina Dickson, Lynette McLeod and Vivek Nemane; and Ph.D. Innovation candidates Darren Marshall, Roxane Blackley and Lisa Yorkston. Dr. Kylie Lingard provided input to early drafts of the practitioner profiles and wild dog case studies.

Professor Theodore Alter is grateful for the support of his research assistants Alyssa Gurklis and Gretchen Seigworth who provided organisational and administrative support, and intellectual contributions during the course of this work. We also thank Alex Riviera and Jeffrey Bridger of Pennsylvania State University for their contributions to the theoretical framing of this book and analysis of the practitioner profiles.

A special acknowledgement goes to Madison Miller of Pennsylvania State University who, in addition to leading Chap. 2, provided valuable suggestions and substantive, concrete revisions to the entire manuscript. Thank you, Madison, for your enthusiasm for the work, your collegiality and professionalism, and your commitment to this shared enterprise of narrative enquiry.

Many thanks to Nick Melchior at Springer Publishing for his initial enthusiasm and support for this project, and to Ilaria Walker for her ongoing encouragement.

Many government, industry and community partners involved in this research program provided support, insight and encouragement. We thank them and all of the individuals interviewed for this book for trusting us with their personal and professional perspectives on the challenging issue of community pest management.

Contents

1 The Context of Community Pest Management in Australia: Myths, Stories and Narrative Enquiry 1
  1.1   The Current Context of Pest Species Management 3
  1.2   Myths and Stories: The Foundation of Pest Management in Australia 6
        1.2.1   Terra Nullius: Nobody's Land.................... 6
        1.2.2   No Place like Home: Acclimatisation 7
        1.2.3   Building a Nation: Nationalism................. 7
        1.2.4   Patriotism: A New Narrative................... 8
  1.3   The Link Between Culture and Community Pest Management...................................... 9
  1.4   Community Pest Management: Breaking It Down........... 11
  1.5   Concluding Comments 14
  References 17

2 Developing and Using Narratives in Community-Based Research 21
  2.1   What Is Narrative and What Role Does It Play in Our Lives?...................................... 22
        2.1.1   Narrative Across Disciplines 24
  2.2   Why Is Narrative Important in Addressing Community Issues?...................................... 27
        2.2.1   Framing the Issue Through Narrative.............. 28
        2.2.2   Community Narratives 29
  2.3   Why Use Narrative for Community Engagement in Pest Management? 30
        2.3.1   Narrative Enquiry as a Research Methodology 30
        2.3.2   Narrative and This Study 31
        2.3.3   Approaching the Practice Profiles and Case Studies 32
  References 33

Part I Practitioner Profiles: First-Person Practice Stories

3   **Profile Introduction and Analysis** 39
   3.1   Introduction 39
   3.2   Interpretation and Analysis 41
   Reference 42

4   **Practitioner Profile (Lisa Adams): "We Cannot Carry
   the Whole on Our Own—We Have to Work Together"** 43

5   **Practitioner Profile (Ben Allen): "If People Don't
   Want to Do It, It's Not Going to Change Anything,
   so I Work with People"** 51

6   **Practitioner Profile (Dave Berman): "Building Trust
   with Community Members"** 57
   References 65

7   **Practitioner Profile (Brett Carlsson): "The Dogs Are There
   and the Tools Are There—We Just Need to Work Out
   the People."** 67

8   **Practitioner Profile (Barry Davies): "You've Got
   to Personalise It"** 73

9   **Practitioner Profile (Peter Fleming): "What's in It
   for the Stakeholder?"** 81

10 **Practitioner Profile (Matt Gentle): "People Are Key
   to the Solution."** 87

11 **Practitioner Profile (Jess Marsh): "It Takes Years to Build
   Trust and a Day for It to Go"** 93

12 **Practitioner Profile (Darren Marshall): "A Learning Journey
   for Me and the Community"** 99

13 **Practitioner Profile (Greg Mifsud): "Pest Management
   Is All About People"** 109

14 **Practitioner Profile (Mike Reid): "It's not Perfect,
   It's Complex—And That's Ok"** 119
   References 127

15 **Practitioner Profile (Harley West): "Whatever They Say,
   I Treat It as a Serious Question"** 129

16 **What Can We Learn from the Practitioner Profiles?** 135
   16.1   The Meta-Lesson: The Struggle for Change 136
   16.2   Practice Lessons 139

16.2.1   Developing an Engagement Practice—Reframing
         "Community"... 141
16.2.2   Developing an Engagement Practice—Look for
         Opportunities to Learn 142
16.2.3   Developing an Engagement Practice—Learning to
         Listen 144
16.2.4   Developing an Engagement Practice—The Drive to
         Make a Difference 146
16.2.5   Developing an Engagement Practice—The Practical
         Side of the Business.......................... 148
16.2.6   Developing an Engagement Practice—There's Only
         so Much You Can Do 151
16.2.7   Developing an Engagement Practice—Combining
         Theory and Practice 152
16.3   Learning from Practice Stories....................... 154
Reference... 155

Part II   Wild Dog Groups—Three Case Studies

17   Introduction: Wild Dog Management Groups................. 159
     References 165

**18   Case Study: Mount Mee Wild Dog Program—Moreton Bay
     Shire, Queensland** 167
     18.1   Case Study Context 167
            18.1.1   Geographic and Physical Context 167
            18.1.2   Wild Dog Management Context 169
     18.2   "The Community Won't Be Ignored": A Narrative of Local
            Government and Community Action Compiled from 12
            In-Depth Interviews 171
     18.3   Narrative Analysis—Mount Mee 176
     References 178

**19   Case Study: Ensay and Swifts Creek Wild Dog Groups—East
     Gippsland, Victoria**...................................... 179
     19.1   Case Study Context 179
            19.1.1   Geographic and Physical Context 179
            19.1.2   Wild Dog Management Context 181
     19.2   "It's Easy to Feed a Farmer's Frustration": A Narrative
            of Two Community Responses Compiled from 17
            In-Depth Interviews 183

	19.2.1	Ensay	184
	19.2.2	Swifts Creek	185
	19.2.3	Policy Action	186
19.3	Narrative Analysis—Ensay and Swifts Creek		188
	19.3.1	Ensay	189
	19.3.2	Swifts Creek	190
References			191

20 Northern Mallee Declared Species Group—Esperance, Western Australia ... 193
    20.1 Case Study Context ... 193
        20.1.1 Geographic and Physical Context ... 193
        20.1.2 Wild Dog Management Context ... 195
    20.2 "Singing from the Same Hymn Sheet": A Narrative of Wild Dog Group Action Compiled from 13 In-depth Interviews ... 197
    20.3 Narrative Analysis—Northern Mallee Declared Species Group (NMDSG) ... 202
    References ... 204

21 Three Wild Dog Group Case Studies: A Meta-analysis ... 207
    21.1 Emotional Dimensions ... 208
    21.2 Capacity to Act ... 209
    21.3 Leadership and Community Structure ... 211
    21.4 Power and Influence ... 212
    21.5 "Naming and Framing" Issues in Wild Dog Management ... 214
    21.6 Conclusion ... 216
    References ... 217

Part III   Learning from Stories of Practice

22 Conclusions ... 221
    References ... 231

Glossary ... 233

Chapter 1
The Context of Community Pest Management in Australia: Myths, Stories and Narrative Enquiry

Abstract In this chapter, we introduce concepts that ground our research purpose, methodology and analysis. We provide key background information to illustrate the current ecological and social landscape of pest management in Australia. This includes:

- Outlining how Australia's history and culture influence the current context of pest management today.
- Presenting our framing of pest management as a community problem that requires collective action.

We make recommendations about how these concepts might help practitioners in their work with community members. These include:

- Being alert to the influence of power dynamics in shaping community interactions, including how sources of knowledge are used and valued.
- Developing skills in critical reflection about community values and beliefs as a necessary foundation for addressing public issues such as pest management.
- Recognising that terms and ideas, such as "community engagement" and "shared responsibility", can carry diverse meanings for different people. Practitioners should push past assumptions to define terms with community members within their specific context.
- Suggesting that stories are one way to build common purpose and increase the likelihood of shared community action.

This chapter aims to persuade you, the reader, that best practice community pest management is achieved through a balance of both social and scientific knowledge. We suggest that:

- Storytelling and dialogue allow individuals and groups to rethink present issues and future possibilities, encompassing diverse or less dominant perspectives to challenge current thinking.
- Readers can use the stories to prompt reflection about ways of approaching issues in working with people in community.

© Springer Nature Singapore Pte Ltd. 2019
T. M. Howard et al., *Community Pest Management in Practice*,
https://doi.org/10.1007/978-981-13-2742-1_1

We Begin with a Story

> The rabbits came many grandparents ago… The rabbits spread across the country. No mountain could stop them; no desert, no river. Still more of them came… Rabbits, rabbits, rabbits. Millions and millions of rabbits. Everywhere we look there are rabbits. (Marsden 1998)

In his picture book collaboration with artist Shaun Tan, author John Marsden uses the widespread and destructive nature of the rabbit in Australia as an allegory for the European colonisation of Australia. Although the European rabbit invasion of Australia and the associated environmental and social consequences have been the subject of other scientific and social investigations, those stories have largely avoided the colonial parallel that Marsden so clearly draws. This parallel is important as it pushes land managers and citizens alike to recognise how history impacts species management today. Marsden's rabbits invade the country, overwhelming the original inhabitants and changing their environment forever. This example illustrates the way narratives, or stories, can shape our understanding of pest management. It sets the scene for the practitioner profiles and community narratives that are at the heart of this book.

In this chapter, we consider the power inherent in "story"—the many different ways it can be told and read and the different lessons it might contain. Through a broad sketch, we consider some of the powerful foundation myths that influence stories of pest management in Australia. We have drawn on a wide range of concepts and sources, including history, literature, public policy and theories of community development, sociology and political economy. In this chapter, we share some of these sources so you can understand the uncertainties and ambiguities that we have encountered in our work and how this has framed our enquiry. While there have been many valuable and popular books written about the threats posed by pest species in Australia, our collection keeps the focus firmly on the *human* species. We are interested in the way that humans respond to the threat posed by widespread and established pests, and this curiosity has been the stimulus for our narrative enquiry.

The narratives collected in this book explore individual and collective perceptions of current practices, obstacles and opportunities related to community pest management in Australia. Storytelling is an imperfect art; it cannot be authoritative or absolute, because it is always possible to retell the story and therefore change it. The reader, through their interpretation of the story, becomes a co-creator. In the act of telling a story, hearing a story, capturing a story and reading a story, a range of possible "versions" can emerge. This does not dilute the value or power of stories. Their ability to generate empathy, transmit information and sustain social relationships is well documented as part of a powerful human tradition of knowledge sharing and meaning making (Goodson et al. 2010). Stories can be simple but they can also be subtle and complex. The stories collected here reveal key themes about the nature of agricultural production in Australia, such as economic pressures, geographic conditions and governance arrangements, while also surfacing deeper influences of culture and social change. They provide insights into the philosophies and practices that underpin efforts to mobilise citizens, both individually and collectively, to address difficult landscape management issues.

1.1   The Current Context of Pest Species Management

Australia's landscape is crowded with evidence of pest species. Many of these species are widespread and established, defeating efforts to remove or eradicate. Animals, plants, aquatic species and insects—both introduced and native—are known to be invasive and destructive to biodiversity and agricultural productivity. Rabbits, camels, cane toads, carp and fire ants, blackberry, skeleton weed and athel pine are just some of the species that are designated as pests in Australia. Many of these pests were introduced during European settlement for food or hunting purposes. Others have been released as biological controls or for economic development. Sometimes, they are native or naturalised species that have become invasive, as landscape conditions have changed or they have been introduced to other parts of the country. Despite strict biosecurity regulations, international trade and travel continue to spread pests between countries and within the borders of Australia.

> Pest species can be viewed through a range of different lenses.

These include the biophysical sciences and technical fixes that have dominated pest species management in recent times (Braysher 2017); the growing awareness of the key role that human behaviour plays in driving landscape change and management (Ford-Thompson et al. 2012); and the way that these realms are represented and shaped by decision-making frameworks such as legislation, policy, best practice advice and culture (Coleman et al. 2017). The impact of pest species can be ecologically catastrophic; we see this in the widespread erosion and soil degradation caused by rabbits or in the loss of native species due to competition from introduced species, such as carp in Australian waterways or cane toad impacts on small mammals in northern Australia. The impact can also be emotionally charged, as concerned citizens witness drastic changes to flora and fauna that may never be reversed. Many excellent histories document the challenges inherent in trying to control established pests, with rabbits receiving much attention throughout the decades due to their widespread and devastating impact (Rolls 1983; Coman 1999). These accounts include detail about the economic and social impacts experienced by those affected by widespread pests, and chart the varying attempts of public policy to address these persistent species. The conclusion is generally that "despite concerted efforts over many decades … the species that were considered to be established pests in the 1900s are still pests today" (Braysher 2017, p. 11).

Designing and implementing pest species management requires a balance between public and private values. Widespread and established pests impact negatively on public-good values such as biodiversity and ecosystem function; they compromise the regenerative integrity of the landscape and they reduce amenity through the introduction of prickles, spikes, allergens and other unwelcome features. A loss of genetic diversity reduces the pool of natural resources available for scientific exploration and can lead to homogenous landscapes with limited variety of plants and animals. Pests also cause negative financial and environmental impacts for private agriculture, increasing reliance on herbicides and insecticides, and in some cases reducing the economic viability of certain enterprises. Other social values include animal welfare concerns which may challenge management activities considered inhumane or unnecessary by those defending animal rights. Determining where the responsibility lies for managing established pests requires a careful balancing of these public and private interests. This balance is continually renegotiated as impacts of pests wax and wane.

Pest control in Australia has been through many phases. Attempts to address widespread pests such as rabbits, wild dogs and foxes have included financial rewards (bounty systems), different technologies (poisons, trapping, barrier fencing) and large-scale culling and control programs. Leadership in pest control has, at different times, been provided by individual landholders, industry groups and government agencies. The early days of European settlement were characterised by a spread of exotic and introduced pests, either deliberately cultivated or accidentally released through trade routes and transports. Tim Low records what might be the first policy determination regarding pest species, as pigs brought to Australia by the First Fleet escaped from their pens and devoured "precious rations" at a vulnerable time for the new settlement "the order was soon given that trespassing hogs could be killed" (Low 1999, p. 25).

More recently, the management of widely established pest species in Australia has become organised around a policy principle of *shared responsibility*. This principle reflects an economic model of cost-effective control known as the "generalised invasion curve". This curve charts the ever-diminishing returns on efforts to control or manage those pests which have become entrenched in the Australian landscape. The curve first appeared in the Victorian government's Invasive Plants and Animals Policy Framework (Agriculture Victoria 2015) and has since become influential in the development of other state and federal government strategies. The curve clearly identifies four stages of pest management. These include:

- pre-introduction "prevention", the phase associated with rigorous biosecurity controls and effective monitoring of borders;

- the post-introduction "eradication" phase, when species are still in small numbers and highly localised populations, increasing the chances of removing the problem before it becomes widely established;
- once eradication is deemed unfeasible due to widespread establishment, the curve identifies "containment" as a way to deal with rapidly increasing distribution and abundance of pest species.

The final phase is described as "asset-based protection", which acknowledges that the pest species is "widespread and abundant throughout its potential range" (Agriculture Victoria 2015). This phase is generally associated with reduced government investment and coordination of pest control, and increasing emphasis on individual and collective landholder responsibilities.

The widespread adoption of the invasion curve has practical consequences for the human dimensions of pest management. While it presents a pragmatic economic argument, it excludes other values that may be important in deciding the balance between public and private interests, where investment and efforts should go, what assets should be protected and who bears the costs of managing these widespread pest species. The Australian Pest Animal Strategy 2017–2027 and Australian Weeds Strategy 2017–2027 affirm "shared responsibility" as the preferred model for reducing government investment (in line with the assumptions of the invasion curve) and increasing the role of "people and organisations from the local to the national scale". These strategies emphasise provision of information, raising awareness and collecting good data as steps on the pathway to achieving this increase in community management.

These techniques can be categorised as "extension services", a form of community engagement which focuses on knowledge transfer and providing expertise through networks of affected and interested landholders (Barr 2011).

Extension services are usually driven by a particular agenda such as increased adoption of new technology or management techniques, and have been historically funded by either government agencies or industry bodies (Wilkinson 2011). The extension approach has been critiqued for promoting a "one-size-fits-all" approach to community engagement which elevates the role of expert advice over the needs and knowledge of the affected land manager (Stobbelaar et al. 2009), missing opportunities to engage passion and motivation through an over-reliance on technocratic solutions (Peters et al. 2010). Most importantly, in the context of shared responsibility, this extension model may have encouraged an over-reliance on institutions such as government and industry bodies to determine and drive pest management efforts. Although the language of "shared responsibility" is being implemented in different jurisdictions across Australia as a legislated citizen biosecurity obligation, research suggests the concept is poorly understood by community members (Craik et al. 2012) who continue to look to "experts" for the answers to the pervasive problem of pest management. The history of widespread and established pest management in Australia illustrates that, just as no one stakeholder "owns" the pest problem, no one stakeholder can be responsible for finding the answers.

1.2   Myths and Stories: The Foundation of Pest Management in Australia

This is a book about stories and the power of narratives to shape our understanding and actions. Some stories, told and re-told over time, assume mythical status. They persist because they represent a widely accepted version of reality and help us explain and understand our experience. Donald Horne, the Australian journalist and cultural theorist, wrote that each society has legitimating myths which have the "magical quality of transforming complex events and actions into simple meanings that explain and justify action … and have the transforming effect of hiding the contradictions and inadequacies of reality" (Horne 1985, pp. 172–173).

To enrich our understanding of the challenges faced in contemporary community pest management, we begin with a brief consideration of the myths that have contributed to shaping Australia's national identity and the community response to pest species over time.

1.2.1   Terra Nullius: Nobody's Land

In Australia, rural landscapes hold stories of economic prosperity, innovation, ecological damage and cultural alienation. The spread of pests followed, and sometimes preceded, the arrival of European settlers by boat in 1788. The number of weeds established around the first settlement at Sydney Harbour was recorded at around twenty in just over a decade, and weed populations continued to increase until the landscape came to resemble European habitats. This "ecological insurrection" (Low 1999) continued across the continent as settlers set out to claim territory for the British empire. The displacement of native flora and fauna by introduced species occurred in parallel with the colonial displacement of Australia's indigenous peoples, whose sovereignty was dismissed through a false narrative that the continent was uninhabited. The evidence for this claim of *terra nullius,* or "nobody's land", included the apparent lack of agricultural cultivation in the mould of European agrarian practices. From initial settlement in 1788, Aboriginal farming and ecological knowledge systems were destroyed as access to traditional trade and travel routes were fenced off by the new owners. Disease, forced removal and state-endorsed violence saw the Aboriginal population drastically reduced as the British colony spread across the country.

Since the myth of *terra nullius* was overturned in 1993 [Mabo v Queensland (No. 2) (1992)], concerted efforts by historians, archaeologists, anthropologists and Aboriginal and Torres Strait Islander people have begun to restore knowledge of farming, fishing and land management practices and how they shaped the Australian landscape over millennia (Gammage 2012; Pascoe 2014). Sadly, this retrospective recognition cannot address the persistent biophysical consequences of settlement and the damage caused to the environment by the introduction of pest species. These scars remain visible on the ancient skin of Australia's landscape. They also run deep in the psyche of the country and its people.

1.2.2   No Place like Home: Acclimatisation

As colonisation intensified, a new narrative emerged that encouraged settlers to introduce a range of non-native plants and animals in order to recreate the longed-for landscapes of "home". Reinforcing the myth of "no one's land", acclimatisation advocates of the 1800s persuasively argued for the introduction of a range of species that would populate the apparently "unfurnished" plains of Australia with familiar, and therefore superior breeds. "How delightful, of a summer's evening, to hear the English blackbird and in the morning the English skylark?" suggested Mr. Francis to his fellow settlers in 1862. "I should like to see again my old impudent friend the sparrow, and the robin and the wren; even the cuckoo would remind us of merry England; and a run for few miles over our broken mountains after a wily fox would be no bad sport" (Francis 1862, p. 10). In these words, we hear a romantic, sentimental and somewhat tragic paean to the landscapes, flora and fauna of the Old World. These pioneers of plant and animal introduction seemed to be motivated by both the project of colonial domination and a providential drive to continue God's work in furnishing the earth more completely. These imperatives combined to reinforce the apparent inevitable and justified settler-driven domination of the country, the indigenous inhabitants and the plants and animals of Australia.

1.2.3   Building a Nation: Nationalism

As settlers put down roots and developed economic and political infrastructure, there was a growing need for myths to anchor them to the Australian landscape with a sense of belonging distinct from their colonial origins. Epic poems such as Banjo Patterson's "The Man from Snowy River" and Henry Lawson's "The Drover's Wife" elevated agricultural landscapes as the authentic backdrop of white Australia. These stories were populated by stereotypical pioneers, "tough men [and women] created by the tough Australian bush" (Clendinnen 2006). Dorothea Mackellar, in her iconic poem "My Country", famously described the landscape as the "core of my heart" and her evocation of "a sunburnt country" valorised the challenges faced by pioneering farmers who struggled to dominate the "wilful lavish land" of the Australian bush. These lyrical imaginings complemented religious suggestions that occupation and settlement were in the mould of a Christian journey to the "Promised Land" (Veracini 2010).

The spread of pastoralism across the country elevated one industry above all others in the national consciousness as a representation of colonial providence and success. The sheep industry became regarded as the economic powerhouse of agricultural development. A colonial attachment to the Old World supported this romantic view, which was reproduced and reinforced by imagery of shepherds and their flocks bathed in the golden light of early landscape painting in the New World. This sentimental representation is still powerful today and rests on the mythical claim that Australia's colonial wealth was made "off the sheep's back", resulting in "a particular culture of pastoralism and agriculture on the land as symbolic of the country's national identity" (Rangan et al. 2014, p. 119). This ideal of the sheep industry has been persistent, despite the challenges posed by wild dog attacks, economic pressures and the realisation of the central role that government subsidisation has played in maintaining the industry over time (Seddon 2003).

1.2.4 Patriotism: A New Narrative

Artists were at the forefront of efforts to develop a link between this powerful vision of the pastoral ideal, a growing love of Australian nature, and patriotism in the public consciousness. Pioneers such as Alec Chisholm (1890–1977) and May Gibbs saw a unique Australian identity in the flowers and fauna of the bush, and suggested that by preserving and encouraging these species the new settler nation would become more authentically grounded and not so precariously transplanted, easily dislodged and vulnerable to being overrun. Scientists with a passion for communication and the unique ecology of Australia began to bang the drum more loudly in the latter half of the twentieth century, calling out to the general public about the destructive impact of pest species in Australia and invoking the patriotic notion that "any lasting notion of Australian nationhood must arise from an intimate understanding of Australia ecosystems" (Flannery 2005, p. 390). In his important book *The Future Eaters,* Tim Flannery reminds us that the history of colonisation and settlement are still influential in the land management practices of the present. He writes that "the problem of cultural maladaptation seems to be particularly acute in Australia ... [because] the land does not hold [its people] comfortably" (Flannery 2005, pp. 389–390).

1.3   The Link Between Culture and Community Pest Management

Acknowledging the key role of culture in driving our understanding of the environment and society encourages us to think more about how "pest" species are defined. The definition or categorisation of particular animals or weeds as pest species can reveal a lot of information about the social context of the issue, as do the practices of coping with perceived overflows [excesses of species] (Hillier 2017). By now the reader will be aware that our interest in stories is linked to our belief that *practical decisions about ecological restoration are entwined with culturally driven assumptions and judgements* (Trigger et al. 2008). Pest species are generally defined as not native (or alien) to a particular ecosystem. The degree to which organisms are described as "invasive" or "feral" or "pest" is linked to the specific environmental or economic values that the affected community holds dear (Wallis and Ison 2011). Domestication or cultivation of particular species demonstrates human ownership and control, and leads to important conceptual distinctions being made between acceptable and non-acceptable populations. For example, domesticated populations of pigs are never referred to as "invasive" or "introduced", while wild pigs are known as "feral hogs" and are the focus of intensive control efforts through trapping, baiting and shooting. The distinction "depends largely on human values ... by attempting to manage invasive species, we are affirming our economic and environmental values" (Beck et al. 2008, p. 419).

Pest species are not always "outsiders" that have been introduced to an unprepared ecosystem. They can also be native species that have become problematic due to translocation or through species expansion as a result of changing land management practices. For example, despite their place on Australia's coat of arms, both kangaroos and emus can be described as "pests" because of their impact on certain agricultural interests. Kangaroos compete for good grazing country and easily jump the fences that control domesticated livestock. Emus are accused of migrating en masse during times of drought and causing havoc for grain crops. Deciding when and how to manage these species is controversial and highlights the role of human values in determining the response.

This relationship between "native" and "alien" species is a recurring theme in Australian pest management that may be linked to anxieties about land management or agricultural practices in a recently settled landscape. This anxiety ebbs and flows and is reflected in changing policy responses and political discourses.

As we have seen, plants and animals were an important part of the colonial project, with acclimatisation and naturalisation both strategies for spreading evidence of new ownership of Australia (Rangan et al. 2014, p. 118). This interpretation of "invasion biology" [as part of an] "imperial ecology" of domination and acculturation (Caluya

2014, p. 31) complicates the technical or scientific perspective. It is discomforting. However, it also encourages us to investigate the central role of culture in our understanding of the relationship between people and pests. This is why we look at stories that emerge from the culture of pest management and seek to understand how they relate to the task of motivating community action.

Addressing the Royal Colonial Institute in 1875, acclimatisation advocate Edward Wilson advised his audience that eradication of "problem" species must be supported where it was deemed necessary in order to "assert our right to destroy some things for the purpose of smoothing the path of more valuable things" (Wilson 1875, p. 19). In discerning between *valued* and *expendable* or *disposable* species, Mr. Wilson and his fellow acclimatisers articulated a tension that remains a challenge for community pest management today. Pests bring a complex web of emotional connections, cultural beliefs and even religious attachments to challenge obvious and effective control measures. Control of feral donkeys and camels in Central Australia has been complicated by the biblical association of both species, leading Aboriginal communities to question the moral and ethical dimensions of reduction efforts (Vaarzon-Morel and Edwards 2012). Families with pet rabbits respond fearfully to the spread of rabbit control viruses, making an emotional distinction between the domestic and "wild" rabbit. In rare cases, the visibility and prolific impact of a species will prompt a coordinated response across the community (Fitzgerald et al. 2007). For example, the ugly face and skin of the cane toad undermines sympathy for this introduced species and encourages questionable behaviour such as "toad golf" and other inhumane responses. However, for many other pests, the challenge of developing effective management solutions is complicated by deep and often unaddressed cultural dimensions.

Many of the species that were enthusiastically promoted as "suitable" for acclimatisation are now established and widespread pests in Australia, such as foxes and deer. The transition from a valued species to a problem species is hard to map and can be difficult for individual citizens to understand and accept, unless they are directly affected by the negative or destructive impacts of the species. Pest animal control is often challenged by public concern about humane treatment of animals, raising another important cultural dimension of this problem. Attempts to control species such as wild horses or kangaroos through shooting prompt a visceral response from community members who have an emotional connection to the animals. Opponents reject the scientific or production knowledge that informs the control recommendations and are often successful in preventing any control from going ahead. These situations highlight how assumptions about what is valued and worthy of protection are influential in driving the human response. This confounding subjectivity of human understanding is a feature of community engagement in complex issues.

The wild dog–dingo hybridisation debate provides a compelling illustration of this confusion between native and non-native, valued and non-valued, pest and pet. Where wild dogs (*Canis lupus familiaris*) are seen as "pests", the dingo (*Canis lupus dingo*) retains a special status as a native dog. Uncertainty about species interaction between dingoes and wild dogs has led to a confusing mosaic of different legislation across Australia, where dingoes are categorised as either a "protected species" or a "declared pest" depending on the jurisdiction (Purcell 2010). Wild dog management can raise concerns about the survival of the dingo, particularly in regards to its role in the cultural and spiritual life of Aboriginal Australia (Rose 2011). It is an iconic animal that for some, also resonates with emotional symbolism of oppression and resistance. The dingo has become trapped between the status of pest animal and totemic creature (Purcell 2010).

> In this sense, the wild dog can be seen to parallel the colonial impact of settlement, spreading across the country and potentially displacing dingo populations. The way the "problem" of wild dogs is framed is a significant feature of how management regimes are formed and what triggers are used to inspire collective action (Howard et al. 2018).

1.4 Community Pest Management: Breaking It Down

Having broadly sketched the narrative underpinnings of pest management in Australia, we now turn to the particular concerns that shaped our investigation and our approach to these challenging questions of pest management, community engagement and shared responsibility. Understanding the historical and current social and cultural context of pest species management helps us think about the challenge of achieving "shared responsibility" as an essentially social and political problem. This research frames the "problem" of community action for pest species management as dynamic, evolving and essentially social.

> Our starting point is that:
> pest management is a *community problem,* which requires *collective action* in order to achieve best results across the landscape; and
> *collective action* requires people to work together to develop a *shared vision* and *commitment,* to the problem and to each other, in order for that action to be sustained over time, in response to the persistent nature of pest species.

This framing agrees that "invasive life … has the power to create communities and supports analysis of the social dynamics set in motion in the context of dealing with invasive life" (Everts 2015, p. 196). The shared recognition, understanding and acceptance of a threat from a pest species can lead to a shared decision to act collectively. As such, a pest species can be a trigger for community formation and action, particularly when there is a collective recognition of the threat, acceptance and a collective response. However, recognition and acceptance of a threat is not sufficient to create this community response (White et al. 2008). A complex social dynamic of community formation underpins the progress from threat awareness to collective action, and to some degree this progression relies on a compulsion to join, belong or accept nomination to a shared community. This book strives to explore these complexities through the practitioner stories and wild dog management group case studies. To understand how communities form and respond to a range of different species, we examine different examples of individual and group responses across a range of geographical landscapes. In this way, our work contributes to the identified need to understand in more detail how community is produced, or not, in the context of pest species (Everts 2015).

The capacity of a community to work together to control a specific pest cannot be examined without unpacking how this collective effort began; who the leaders of the effort may be; what role they play in the collective and in the wider community; and how power is expressed through the actions that are taken (Gaventa 1980). It also requires that we interrogate the context for entrenched advantage and disadvantage, and consider how this might influence the legitimacy of group action. We agree that "how we deal with nature is often a reflection of how we deal with humans and is thus reflective of human politics" (Caluya 2014, p. 35). This reference to the ubiquity of politics introduces an important theoretical frame for our work. Politics is an expression of power. All individuals in society experience power, as they both use their own power and are susceptible to the power wielded by other people and organisations. The narrative enquiry method offers a way to understand the impacts of power in collective community action. As our brief discussion of foundation myths illustrates, a complex interaction between culture and power underpins our attempts to stimulate collective community pest management. Throughout this enquiry, we remain alert to this interaction, particularly where entrenched patterns of power and privilege seem to constrain or distort community efforts to develop locally accepted management approaches (Cornwall 2002).

As scholars, we believe that it is important to probe beneath the surface to reveal the way that entrenched patterns of advantage and disadvantage shape our understanding of community action. Throughout our narrative analysis, we ask: What are the consequences of any particular decision or path or action? And whose interests are being served? (Schmid 2008). We do this by examining the practice stories and case studies for the versions of community engagement they illustrate and the expectations that are attached to these. Grounded in the belief that communities of individuals can be meaningfully involved in determining the form and substance of decisions that will affect them, community engagement has become a common phrase for a wide range of participatory processes (Eversole 2011; Koontz and Thomas 2006). Decision-makers in the public policy arena increasingly describe an *engaged* community as key to the successful design and implementation of regulation and policy (Australian Public Service Commission 2007; Walker 2011). There are many unarticulated assumptions that underpin this position, including that community engagement reliably translates into more acceptable decisions (Adams and Hess 2001) while also delivering better outcomes (Ford-Thompson et al. 2012). Although the study and practice of community engagement is common to many fields of public policy governance and research, it is often over-simplified (Adams and Hess 2001). Sometimes, engagement is described as a set of processes and procedures that can be applied to achieve specific results. Other times people use the word "engagement" to suggest transformational ideals of community-driven social change and power transfer through deliberative participatory processes (Dryzek 2000)—in other words saying that engagement means working *with* people, involving citizens in decisions, and letting their skills and perspectives lead action. The range of possible interpretations suggest that philosophies and practices of engagement are worthy of investigation and reflection, because it helps us understand that "engagement means specific things to specific people, who use the concept to refer to specific processes, situations, and outcomes, contingent on their school of thought and practice" (Fear et al. 2006, p. 50).

> Throughout this book, we reflect on the social, political, economic and cultural undercurrents that characterise particular examples or experiences of community engagement. From our viewpoint, this critical examination is essential to building a professional practice of engagement that allows individuals to meet the specific needs and challenges of their work with community.

Articulating and reflecting on the range of assumptions that accompany community engagement efforts can also serve as a good reality check. For example, critical reflection is necessary to understand the philosophical foundation of a concept such as "shared responsibility", the way it may be interpreted by different audiences and the likely challenges such a concept will encounter. Knowledge formed from reflection can then equip us with important information about the kinds of processes that might best facilitate a "shared responsibility" outcome. Beyond an outcome, critical reflection prompts people to develop and strengthen a professional practice of engagement that moves beyond best practices to become flexible and deeply principled about how to work with people.

For us, the development of a principled practice of community engagement is the key to achieving community-led collective action because this allows each practitioner and community leader to respond authentically, and with integrity, to a range of challenges across a variety of issues. For example, it would be naive to suggest that simply issuing an invitation to 'shared responsibility' will be sufficient to get individual citizens to act, unless they understand this concept as relevant and appropriate in their particular context. A well-developed understanding of community engagement practice will increase the likelihood that community values and beliefs are sought out and explicitly integrated before the invitation to act is issued (Howard 2017). This introduces another important theoretical frame for our work: maintaining an awareness of the possible power imbalances that can result from a primarily technocratic or expert-driven model of knowledge production in regards to public issues such as pest management (Subramaniam 2001).

As we have previously noted, despite growing recognition of the importance of community engagement for pest management, there is still a strong tendency for policymakers, researchers, land managers and landholders to focus attention on biophysical and technical fixes; and this tendency reflects a privileging of particular sources of knowledge. It is grounded in scientific concepts of reliability and rationality that can undermine confidence in community values and local knowledge sources (Fischer 2005). Power is linked with knowledge, and controlling the definition of what is accepted as "real" or "valid" knowledge, particularly in a decision-making context, creates an imbalance that can inhibit participation by non-expert community members (Gaventa 2002). It also reinforces the power of experts to set the agenda. Selecting which piece of knowledge will be shared through media, meeting documents or interpersonal interactions facilitates a power imbalance that can be consciously or unconsciously maintained (Cook 2015). Controlling access to knowledge creates an asymmetry that inhibits genuine participation, and can lead to certain knowledge systems such as traditional knowledge and non-expert views being devalued (Barber et al. 2014; Bridger and Alter 2010).

> Our approach to these problems of knowledge and power in community engagement is to reframe knowledge production as an essentially human endeavour, to which every citizen will have some expertise to contribute. This expertise may be specific to the context of a particular issue or experience, and not broadly generalisable; however, acknowledging and integrating these different types of knowledge is essential in attempts to motivate collective community action. We might describe this as a *democratisation of knowledge,* where community engagement practice aims to build the individual citizen's capacity to not only receive information about the issue under examination but to actively contribute their knowledge.

For individuals to join in collective action, they must feel confident to name and frame issues in ways that are meaningful to them and their context (Lakoff 2010). This may require creating spaces where expert and non-expert knowledge can interact, in order to acknowledge the tensions that inevitably arrive in the definition, exploration and decision-making processes surrounding complex public issues such as community pest management. We begin to see that community pest management is an arena where practices of power sharing, deliberation, knowledge creation and the explicit incorporation of diverse values are necessary in order to build a collective pathway for action (Mathews 2005).

1.5 Concluding Comments

It will by now be clear to the reader that this enquiry is grounded in a research paradigm that recognises and values complexity and uncertainty. We do not shy away from the messy and complex aspects of social research; instead, we embrace a constructivist, or interpretivist, epistemology which allows us to engage with multiple perspectives and explore the resulting tensions. Our research positioning supports a narrative enquiry methodology and the details of this methodology are described in more detail in Chap. 2 and in the empirical sections of the book. This work is firmly focused on stories and the meaning that we make through our telling and retelling.

> Stories, or narratives, tend to change depending on who is telling them and the audience who hears them. They evolve over time in order to remain relevant. Stories are important because they hold clues to both collective and individual experiences, and because they are culturally and geographically situated. In addition, they are also easy to understand, taking advantage of the shared human heritage of oral traditions and experiential learning.

It is important to remember that stories are always subjective. As authors who argue that individual values and beliefs are essential ingredients in understanding experience, we do not claim to "know" the authoritative version. We can only offer a possible interpretation of the narratives that have emerged from our empirical enquiry and point out what we see as meaningful in each account. However, we are aware that this mapping of the narrative experience may not resonate with other readers who bring different experiences and knowledge to the page. For us, this is an essential component of a narrative enquiry. This approach to exploring human experience does not strive to create an elusive "truth"; instead, we suggest that "truth" will lie with the narrator, the reader and the sense that is made by the creative act of engaging with each other. These are the opportunities and the limitations inherent in this approach to research.

Storytelling and narrative forms hold the potential for "re-storying"—a technique that has been successfully used to recast accepted and dominant accounts to embrace or enable diverse, and less dominant, perspectives (Ollerenshaw and Creswell 2002). There is space in this collection for reinterpretation, re-storying and possibly, reconciliation of these different interpretations. For example, when the phrase "invasive species" or "pest species" is encountered, readers are encouraged to interpret this how they wish. For those with inclinations towards rationalist viewpoints of objectivity and scientific positivism, it is standard to think about the invasive species as the "other": the plant, animal or insect that causes damage through a dispassionate lens of remoteness. For those who find themselves experiencing a reflective twinge at the conventional use of the phrase, it is permissible and indeed encouraged, to turn that sense of otherness around—to take the lens and examine the human as part of the biophysical expression of invasiveness or pest behaviour. This perspective can be useful for challenging the assumed dominance of human values in the biosphere and also for extending the possible field of action to include reducing human impacts on biodiversity and earth function.

This collection is essentially concerned with the thorny issue of engaging citizens to collectively address a shared problem. Achieving this requires skills in participatory practices such as problem identification, recognition of different worldviews, conflict resolution and the ability to develop an acceptable shared destination.

Our audience is the practitioner who works with landholders and rural communities to implement collective action; the policymaker and decision shaper within government agencies, industry bodies and non-government organisations; and the individual citizen who hopes to encourage and sustain community action. Some chapters and some stories might appeal and apply more to one audience member than another; yet, as a collective, this book is for all leaders in pest management, because all citizens are decision-makers. We encourage each reader to critically examine the empirical and theoretical dimensions of this work and draw their own conclusions about how it might inform their personal or professional efforts in community pest management. We suggest that developing the skills and awareness to undertake this reflective process is a vital part of facilitating community pest management.

In this work, we do not present authoritative versions because our ambition is to reflect the lived experience of those embedded in both work and place.

Passions run high; knowledge is contested; expertise is mediated by assessments of legitimacy, authenticity and trustworthiness. Different interests reveal the personal values that underpin decisions and illustrate how sometimes these differences can undermine collective action. For engagement practitioners and community workers, addressing these value conflicts is a necessary part of the work they do.

Everyone involved in motivating a community response must be prepared to encounter those who think and see the world like them—and those that do not (Forester 2016). Being able to look beneath the obvious is an important skill for those working with community members in challenging public contexts. To gain trust, establish legitimacy and most importantly, achieve positive collective action, we must avoid acting on assumptions. The ability to reflect on what we see and read, and look beneath the obvious for subtle but influential factors will increase the potential to work across difference, find common ground and ultimately, build a shared platform for collective community action.

Omissions are inevitable. We make no claim to be exhaustive but rather to make visible the influences that have shaped our investigations. Readers may disagree or dispute some of our claims—this is to be expected and indeed, encouraged. If this chapter, and the collection of narratives that follow, stimulates new ideas or alternative ways of looking at the role that individuals and communities play in the management of pest species in Australia, we will have succeeded in our ambition to bring a new paradigm to bear on an established, technocratic view of the problem.

* * * * *

We conclude this chapter by returning to John Marsden's story of the rabbits. The inexorable march of the rabbits across the landscape is described in devastating detail:

> They ate our grass. They chopped down our trees. The land is bare and brown and the wind blows empty across the plains. Where is the rich, dark earth, brown and moist? Where is the smell of rain dripping from gum trees? Where are the great billabongs, alive with long-legged birds?

The story ends with a question that is central to our investigation:

> Who will save us from the rabbits?

"We are the ones we have been waiting for" is a phrase attributed to poet June Jordan. Taken up as a call to action, it has become a touchstone for community development theory and practice, and provides a core rationale for this book (Levine 2013). The phrase suggests that power lies with the action of the people and reminds each individual of their latent potential to take action on behalf of their own and the collective interest. Drawing on this phrase, the answer to Marsden's plaintive question "who will save us from the rabbits?" becomes obvious. There is no one who can "save us" from the rabbit, or any other challenge, beyond ourselves—as individuals and as members of a collective community. Pest management is a community problem. It can also be considered as a problem *of* community. Join us now as we set out on a narrative exploration of community and what it means for pest management in the Australian context.

References

Adams, D., & Hess, M. (2001). Community in public policy: Fad or foundation? *Australian Journal of Public Administration, 60*(2), 13–23.

Agriculture Victoria. (2015). *Invasive plants and animals policy framework*. Biosecurity Strategy for Victoria. http://agriculture.vic.gov.au/agriculture/pests-diseases-and-weeds/protecting-victoria-from-pest-animals-and-weeds. Accessed December 12, 2017.

Australian Public Service Commission. (2007). *Tackling Wicked Problems: a public policy perspective*. Contemporary Government Challenges. Canberra, ACT: Commonwealth of Australia.

Barber, M., Jackson, S., Shellberg, J., & Sinnamon, V. (2014). Working knowledge: Characterising collective indigenous, scientific, and local knowledge about the ecology, hydrology and geomorphology of Oriners Station, Cape York Peninsula, Australia. *The Rangeland Journal, 36*(1), 53–66. https://doi.org/10.1071/rj13083.

Barr, N. (2011). I hope you are feeling uncomfortable now: Role conflict and the natural resources extension officer. In D. J. Pannell & F. Vanclay (Eds.), *Changing land management: Adoption of new practices by rural landholders* (pp. 129–139). Melbourne, VIC: CSIRO publishing.

Beck, K. G., Zimmerman, K., Schardt, J. D., Stone, J., Lukens, R. R., Reichard, S., et al. (2008). Invasive species defined in a policy context: Recommendations from the Federal Invasive Species Advisory Committee. *Invasive Plant Science and Management*, 1(4), 414–421. http://dx.doi.org/10.1614/IPSM-08-089.1.

Braysher, M. (2017). *Managing Australia's pest animals*. Canberra, ACT: CSIRO.

Bridger, J. C., & Alter, T. R. (2010). Public sociology, public scholarship and community development. *Community Development, 41*(4), 405–416.

Caluya, G. (2014). Fragments for a postcolonial critique of the Anthropocene: Invasion biology and environmental security. In J. Frawley & I. McCalman (Eds.), *Rethinking invasion ecologies from the environmental humanities* (pp. 31–44). London, UK: Routledge.

Clendinnen, I. (2006). The history question: Who owns the past? *Quarterly Essay, 23,* 68.

Coleman, M. J., Sindel, B. M., & Stayner, R. A. (2017). Effectiveness of best practice management guides for improving invasive species management: A review. *The Rangeland Journal, 39*(1), 39–48. https://doi.org/10.1071/rj16087.

Coman, B. J. (1999). *Tooth & nail: The story of the rabbit in Australia*. Melbourne, VIC: Text Publishing.

Cook, J. J. (2015). Who's pulling the fracking strings? Power, collaboration and Colorado fracking policy. *Environmental Policy and Governance, 25*(6), 373–385. https://doi.org/10.1002/eet.1680.

Cornwall, A. (2002). Locating citizen participation. *Institute of Development Studies Bulletin, 33*(2), i–x. https://doi.org/10.1111/j.1759-5436.2002.tb00016.x.

Craik, W., Palmer D., & Sheldrake R. (2012). *Intergovernmental agreement on biosecurity review draft report*. Canberra, ACT: Council of Australian Governments. http://www.agriculture.gov.au/biosecurity/partnerships/nbc/intergovernmental-agreement-on-biosecurity/igabreview/igab-draft-report. Accessed December 12, 2017.

Dryzek, J. S. (2000). Liberal democracy and the critical alternative. *Deliberative democracy and beyond: liberals, critics, contestations* (pp. 8–30). Oxford UK: Oxford University Press.

Eversole, R. (2011). Community agency and community engagement: Re-theorising participation in governance. *Journal of Public Policy, 31*(1), 51–71.

Everts, J. (2015). Invasive life, communities of practice, and communities of fate. *GeografiskaAnnaler: Series B, Human Geography, 97*(2), 195–208.

Fear, F. A., Rosaen, C. L., Bawden, R. J., & Foster-Fishman, P. G. (2006). *Coming to critical engagement: An authoethnographic exploration*. Maryland, USA: University Press of America.

Fischer, F. (2005). *Citizens, experts and the environment: the politics of local knowledge* (1st ed.). Durham, NC: Duke University Press.

Fitzgerald, G., Fitzgerald, N., & Davidson, C. (2007). *Public attitudes towards invasive animals and their impacts*. Literature review. Invasive Animals Cooperative Research Centre. https://www.pestsmart.org.au/public-attitudes-towards-invasive-animals-and-their-impacts/. Accessed December 12, 2017.

Flannery, T. F. (2005). *The future eaters: An ecological history of the Australasian lands and people* (New ed.). Sydney, NSW: Reed New Holland.

Ford-Thompson, A., Snell, C., Saunders, G., & White, P. (2012). Stakeholder participation in management of invasive vertebrates. *Conservation Biology, 26*(2), 345–356. https://doi.org/10.1111/j.1523-1739.2011.01819.x.

Forester, J. (2016). The challenge of transformative learning: Mining practice stories to study collaboration and dispute resolution strategies. In R. D. Margerum & C. J. Robinson (Eds.), *The challenges of collaboration in environmental governance* (pp. 338–354). London, UK: Edward Elgar Publishing Limited.

Francis, G. W. (1862). The Acclimatisation of harmless, useful, interesting and ornamental animals and plants. In The Philosophical Society (Ed.), *A paper read before the Philosophical Society*. Adelaide, SA: The Philosophical Society.

Gammage, B. (2012). *The biggest estate on earth: how Aborigines made Australia*. Crows Nest, NSW: Allen and Unwin.

Gaventa, J. (1980). *Power and powerlessness: Quiescence and rebellion in an Appalachian Valley*. Chicago, USA: University of Illinois Press.

Gaventa, J. (2002). Exploring citizenship, participation and accountability. *Institute for Development Studies Bulletin, 33*(2), 1–14.

Goodson, I., Biesta, G., Tedder, M., & Adair, N. (2010). *Narrative learning*. Oxford, UK: Routledge.

Hillier, J. (2017). No place to go? Management of non-human animal overflows in Australia. *European Management Journal, 25*(6), 712–721. https://doi.org/10.1016/j.emj.2017.02.004.

Horne, D. (1985). Who rules Australia? *Daedalus, 114*(1), 171–196.

Howard, T. M. (2017). "Raising the bar": The role of institutional frameworks for community engagement in Australian natural resource governance. *Journal of Rural Studies, 49*, 78–91. http://dx.doi.org/10.1016/j.jrurstud.2016.11.011.

Howard, T. M., Thompson, L. J., Frumento, P. Z., & Alter, T. (2018). Wild dog management in Australia: an interactional approach to case studies of community-led action. *Human Dimensions of Wildlife, 23*(3). https://doi.org/10.1080/10871209.2017.1414337.

Koontz, T. M., & Thomas, C. W. (2006). What do we know and need to know about the environmental outcomes of collaborative management? *Public Administration Review, 66*(S1), 111–121.

Lakoff, G. (2010). Why it matters how we frame the environment. *Environmental Communication: A Journal of Nature and Culture, 4*(1), 70–81. http://dx.doi.org/10.1080/17524030903529749.

Levine, P. (2013). *We are the ones we have been waiting for: The promise of civic renewal in America*. Oxford, UK: Oxford University Press.

Low, T. (1999). *Feral future*. Ringwood, VIC: Viking.

Mabo v Queensland (No 2) (1992). 175 CLR 1.

Marsden, J. (1998). *The rabbits*. Port Melbourne, VIC: Lothian Books.

Mathews, D. (2005). The politics of self-rule: Six public practices. *Connections*, 4–6. Dayton, OH: Kettering Foundation Press.

Ollerenshaw, J. A., & Creswell, J. W. (2002). Narrative research: A comparison of two restorying data analysis approaches. *Qualitative Inquiry, 8*(3), 329–347.

Pascoe, B. (2014). *Dark emu: Black seeds: Agriculture or accident?*. Broome, WA: Magabala Books.

Peters, S., Alter, T. R., & Shaffer, T. (2010). Hot passion and cool judgement: relating reason and emotion in democractic politics. *Connections*, 15–17. Dayton, OH: Kettering Foundation Press.

Purcell, B. (2010). *Dingo*. Canberra, ACT: CSIRO.

Rangan, H., Wilson, A., & Kull, C. (2014). Thorny problems: Industrial pastoralism and managing "country" in northwest Queensland. In J. Frawley & I. McCalman (Eds.), *Rethinking invasion ecologies from the environmental humanities* (pp. 116–134). London, UK: Routledge.

Rolls, E. C. (1983). *They all ran wild: the animals and plants that plague Australia* (Rev ed.). Sydney, NSW: Angus and Robertson.

Rose, D. B. (2011). *Wild dog dreaming*. Charlottesville, USA: University of Virginia Press.

Schmid, A. (2008). *Conflict and cooperation*. Oxford, UK: Blackwell Publishing Ltd.

Seddon, G. (2003). Farewell to arcady: Or getting off the sheep's back. *Thesis Eleven, 74*(1), 35–53. https://doi.org/10.1177/07255136030741004.

Stobbelaar, D., Groot, J., Bishop, C., Hall, J., & Pretty, J. (2009). Internalization of agri-environmental policies and the role of institutions. *Journal of Environmental Management, 90*(2), S175–S184.

Subramaniam, B. (2001). The Aliens Have Landed! Reflections on the rheotoric of biological invasions. *Meridians: feminism, race, transnationalism, 2*(1), 26–40.

Trigger, D., Mulcock, J., Gaynor, A., & Toussaint, Y. (2008). Ecological restoration, cultural preferences and the negotiation of "nativeness" in Australia. *Geoforum, 39*, 1273–1283. https://doi.org/10.1016/j.geoforum.2007.05.010.

Vaarzon-Morel, P., & Edwards, G. (2012). Incorporating Aboriginal people's perceptions of introduced animals in resource management: insights from the feral camel project. *Ecological Management & Restoration, 13*(1), 65–71. https://doi.org/10.1111/j.1442-8903.2011.00619.x.

Veracini, L. (2010). *Settler colonialism: A theoretical overview*. London, UK: Palgrave Macmillan.

Walker, G. (2011). The role for "community" in carbon governance. *Wiley Interdisciplinary Reviews: Climate Change, 2*(5), 777–782.

Wallis, P. J., & Ison, R. L. (2011). Appreciating Institutional complexity in water governance dynamics: A case from the Murray-Darling Basin, Australia. *Water Resources Management, 25*(15), 4081–4097. https://doi.org/10.1007/s11269-011-9880-4.

White, P. C. L., Ford, A. E. S., Clout, M. N., Engeman, R. M., Roy, S., & Saunders, G. (2008). Alien invasive vertebrates in ecosystems: Pattern, process and the social dimension. *Wildlife Research, 35*(3), 171–179. https://doi.org/10.1071/wr08058.

Wilkinson, R. (2011). The many meanings of adoption. In D. J. Pannell & F. Vanclay (Eds.), *Changing land management: Adoption of new practices by rural landholders* (pp. 39–50). Melbourne, VIC: CSIRO Publishing.

Wilson, E. (1875). Acclimatisation: Read before the Royal Colonial Institute. In Royal Colonial Institute (Ed.), *Royal Colonial Institute.* London, UK: Unwin Brothers.

Chapter 2
Developing and Using Narratives in Community-Based Research

Madison Miller and Jeffrey C. Bridger

Abstract Within this chapter, we explore several foundational ideas about using narrative to understand experiences, ourselves and communities, and how to apply these ideas to researching community issues and collective invasive animals management in particular. We learn that:

- Stories help us make sense of the world and our place in it.
- Narratives are relational acts, as narrators place themselves and issues within time and space, within relationships to others, and within larger cultural and institutional narratives.
- Stories have power in our minds and communities, as they impact which actions and outcomes we see as possible and as they allow opportunities for people to come together to coordinate thoughts and actions.
- Narratives can create frames through which people view an issue; and frames can influence the narratives people create about an issue.
- Narrative inquiry is an effective research method for interpreting social experiences. Narratives can provide insights to deepen understandings of complex issues in ways that positivist approaches to science cannot.
- Narrative inquiry is used in this study to reveal tensions and complexity, to offer wisdom and to prompt reflection about approaching community pest management problems. Stories offer ways to focus on approaches rather than pre-prescribed, universal solutions since working with people is fundamentally relational and context-specific.
- Pest management requires collective action, and narratives offer ways for communities to reach shared visions and shared action commitments.
- The narratives in this book are windows into practitioners and communities' experiences that can prompt reflection and offer practical insight useful to people working in communities.

M. Miller (✉) · J. C. Bridger
Department of Agricultural Economics, Sociology, and Education, Penn State University, Pennsylvania, USA

Narrative is an inherent part of the way people understand the world and take action; narratives are not static, but constantly reflecting and influencing our perceptions. This book utilises narrative as a research method and methodology to portray and analyse the stories of pest management practitioners and communities dealing with pest management. The stories tell us about the narrators themselves, the communities in which they work and pest management in Australia more broadly. The profiles and case study narratives help us better understand issues and interactions of community pest management, and how people come together across differences to collectively envision new futures for their communities.

Pest management is one of many examples of a collective action challenge, in which collective effort is required by all members of the community, although individual incentives or rewards may be uneven (Clark et al. 2013; Schmid 2004). Narratives offer a useful tool to address this collective action challenge, because narratives can fuel cooperation by bringing diverse people together for a shared vision that fits within a bigger story of community success (Mayer 2014). In this chapter, we explore the power and prevalence of narrative in shaping individual and community perceptions and fuelling action. We define narrative, including its elements, structure and function. We discuss the importance of narrative and its applications across varying disciplines and the role it plays in shaping both personal perceptions and community issues. Finally, we explore the use of narrative enquiry as a research methodology and how it is used in this book.

2.1   What Is Narrative and What Role Does It Play in Our Lives?

Fisher (1985) calls us *homo narrans*—the species that tells stories. By this, he means in the broadest sense that narrative is how we make sense of the world. Brooks (1984, p. 3) captures the essence of this characterisation when he writes:

> We live immersed in narrative, recounting and reassessing the meaning of our past actions, anticipating the outcomes of our future projects, situating ourselves at the intersection of several stories not yet completed.

We dream in, rear our children with and use stories to both build solidarity and identify outsiders; we learn about society through myth, folklore and legend. As MacIntyre (1985, p. 216) puts it, "man is in his actions and practices, as well as in his fictions, essentially a storytelling animal". Frank (2010, p. 13) goes further in saying that "being human, and especially being social, requires the competence to tell and understand stories". In short, much communication is narrative in form (Bruner 1990) and, precisely for this reason, it is often difficult to recognise the difference between narrative and other types of discourse.

The term "narrative" does not simply designate a type or structure of discourse; narratives are also acts (Smith 1980). Narratives are inherently relational activities, in which the narrator positions themselves, or an issue, within time and space, within relationship to others and within larger cultural and institutional narratives (Clandinin 2013; Daiute 2014). At a minimum, three elements must be present for a narrative to exist. First, the speaker or writer must select the events to describe. This act of selection gives the narrator power, choosing which details or perspectives to include and which to erase in the telling; these become the "events" that make up the narrative. Second, these events must be turned into story elements. Narratives involve the imposition of structure and meaning on the selected events through the use of plot, setting and characters. Third, narratives involve the arrangement of events in space, time and relationship to one another. The narrator decides: which happened first, the chicken or the egg? Where was the chick hatched? And which chicken laid which egg?

Plot weaves events together and reflects the perspective and decisions of the narrator. When the narrator places events into the plot, these individual occurrences become more than a chronicle or serial description of events or actions. The plot brings together the different elements of a story and configures them into a coherent whole. Writers and scholars use the term "employment" to refer to the way narrators weave elements together through the use of plot, which affects how future events both unfold and are understood (Frank 2010). Plots operate in two temporal dimensions: the chronological and non-chronological. Narrators make choices about how to order events along a timeline and how to configure these events so that they form a significant whole (Polkinghorne 1988). In and of itself, a singular occurrence is not particularly meaningful; rather, events take on meaning to the extent that they contribute to the development of the plot (Ricoeur 1984). The process of employment is one of making meaning. The human mind seeks meaning. Stories are mechanisms for creating meaning from external stimuli (Gottschall 2013); communicating meaning and relationships to others (Clandinin 2013); transmitting cultural values (Brown 2005); explaining the unusual and the ordinary (Bruner 1990); and affecting actions shaped by altruistic, ideological and political narratives (Mayer 2014). The very idea of employment assumes that experiences are open to diverse understandings and future possibilities (Frank 2010).

Narrative is also defined by structural patterns. Stories follow common patterns typically involving a complication, crisis and resolution (Gottschall 2013) and must provide closure (some type of endpoint—even if the story continues) (White 1980). The arrangement (where did we start versus where did we end?) and outcome (is the ending happy or sad?) of the structural aspects vary across narratives (Mayer 2014).[1] Culture also greatly affects the dominant narratives available to individuals. As "stories are cultural texts", narratives are useful for revealing differences between

---

[1] Mayer identifies four common narrative plot structures: two of tragedy and two of triumph. In the first tragedy "The Fall", the plot moves from good to bad with the onset of some complicating action, which ends in a bad resolution. In the second tragedy "Dust to Dust", the story starts in a negative state, improves and then ends badly. Conversely, the "Genesis and Exodus" structure involves moving from bad to good with a central period of tension. These "Genesis and Exodus" stories embody the ideal of rags to riches tales. Finally, stories of "Resurrection" start in a positive state with characters subsequently experiencing negative events, which they eventually overcome in a happy ending.

cultures and for prompting reflection about the tensions that might arise from these differences (McAdams 2008 p. 247).

2.1.1 Narrative Across Disciplines

Because of its central role in making meaning of human life, many disciplines have turned towards the use and study of narrative. Narrative enquiry and research is common to the fields of rhetoric and communication studies (Corbett and Eberly 2000). Narrative also has a long and established place in the fields of psychology and sociology (Maines 1993; Leger 2016).

More recently, scholars in political science and organisational management fields have used narratives to understand some of the most puzzling questions about choice and group behaviour. Political science draws on narrative to explain, for example, what is included and excluded in our history textbooks, and offers theories relevant to collective action. Mayer (2014) explains that national narratives, in which the leaders and people of a country are painted as noble and virtuous, leave little room for stories of past transgressions that would detract from the narrative development of the nation over time.

Journalists highlight how politicians use narratives to pursue agendas (Leibovich 2015) or to assert dominance internationally (Garreau 2017). In the business sphere, stories function as mechanisms to share information about the trustworthiness and credibility of coworkers, to make decisions based on past information and future options, to communicate and understand what is valued and rewarded in an organisation's culture and to understand one's place within an organisation (Gabriel 2000).

Narrative is constantly at work within our brains. Insights from the field of psychology guide current understandings of the narrative mind. Gottschall (2013) suggests that humans' storytelling ability is an evolutionary benefit helping us to function better as individuals and as groups. Soon after we begin speaking, we start to tell stories. As we grow, our stories become narratives when we place ourselves in relation to other events, time and space. Before making decisions, we often mentally test potential actions by picturing events as storylines and considering how these stories would or would not fit within the personal narrative we have created. Just as make-believe play prepares children for the challenges of adult life, stories serve as learning tools for adults (Gottschall 2013, p. 58) "project[ing] us into intense simulations of problems that run parallel to those we face in reality". Within this storying process, people decide either to relate to an event or larger issue—or to remain completely separate. Just as narratives allow people to test potential outcomes, larger narratives and personal narratives also influence the actions and outcomes people view as possible. Breaking down former narratives and learning new narratives can help define new possibilities.

Evidence suggests that human brains "catch" emotions from others (Hatfield et al. 1993). When an individual sees or hears descriptions of characters experiencing a given emotion, the part of the brain responsible for controlling and expressing the same emotion is activated (Gottschall 2013). In this way, stories are not only about preparation, but they are also drivers of empathy. Vulnerable to catching the emotions of others through stories, we become susceptible to the power of story to shape our perspectives on issues and also our own personal narratives. Watching a movie, reading a novel or listening to a song often transports our mind out of our present setting and into the character's place—or into another time or place that we have experienced. Stories engross us.

Perhaps this is why, contrary to the quest for objectivity in science and rationality in decision-making, "fiction seems to be more effective at changing beliefs than non-fiction, which is *designed* to persuade through argument and evidence" (Gottschall 2013, p. 150). Fisher (1985) addresses this potential paradox through what he calls "narrative rationality". Unlike scientific rationality focussed on arguments and evidence, narrative rationality operates by offering familiar stories. When people hear or read a story, Fisher (1985) suggests that they look for "narrative fidelity", to see if the story hangs together and makes sense, and for "narrative probability" to test if the story is consistent with other stories they have heard or read and with what they believe to be true in their lives. When these two conditions are met, a story is persuasive or convincing—but not in the way that scientific arguments are persuasive or convincing.

This raises an important point about the contribution that narrative methodology can make to social research. There is an acknowledged tension between traditional scientific (or positivist) approaches to research data, analysis and presentation, and more interpretative (or constructivist) approach. As we explained in Chap. 1, our interest in narrative enquiry emerges from a set of assumptions about the social construction of knowledge that does not seek to meet scientific conventions of replication, generalizability and statistical significance. While narrative accounts cannot provide causal explanations, we believe that they can be the basis for interpretation of a social experience and through a process of narrative accrual, can develop a platform of shared understanding that is particularly relevant to consideration of collective community action (Bruner 1990; Flyvbjerg 2001). Although narrative enquiry and in-depth case studies are accepted qualitative methods, we recognise that there are limitations inherent in this approach, as there are in any research method or paradigm. We selected narrative enquiry in order to deepen our understanding of the complex and sometimes confounding dynamics that may be influential in shaping communication action for pest management.

You may now wonder, are life stories or personal narratives ever true? In response to this question, we ask: what is truth? In this collection, we suggest that searching for absolute truth ignores the truth within each individual's lived experience: derived from a unique set of experiences, conditions and mental processes, and nested within other narratives of place, time and culture. Beyond reconciling the past with the present and future, humans also work tirelessly to construct narratives that provide harmony between personal identity and social norms, power structures and cultural beliefs. McAdams (2008) defines narrative identity as "an individual's internalised, evolving, and integrative story of the self" (p. 242).

It follows that the narratives that we tell about ourselves, our communities and the issues we face are central to addressing complex collective action challenges.

Narratives are not static. Memory operates flexibly in order to align with the emotions of the present, as well as present motivations, goals and social positions (McAdams 2008). There is no absolute truth because "the past is always up for grabs" (Beck 2015). Given that narratives can persuade us to change our stance on or attitude about an issue, we are also constantly constructing narratives about ourselves (Bruner 1990). Mayer (2014) argues that our deepest desire is to know ourselves: "Humans seek not only to understand, not only to find meaning, but also to know who we are and to locate ourselves in the world. To ask, 'Who am I?' is a basic human impulse" (p. 73). Humans constantly and unconsciously filter both positive and negative information into a self-narrative which creates a concept of "self". Professionals who support the use of narrative therapy demonstrate how progressively rewriting one's own personal narrative has shown to improve depression (Adler and King 2012; Leger 2016). People "become themselves through the stories they tell" (Daiute 2014, p. 12). The pest managers and community members you will meet in the following chapters are no exception.

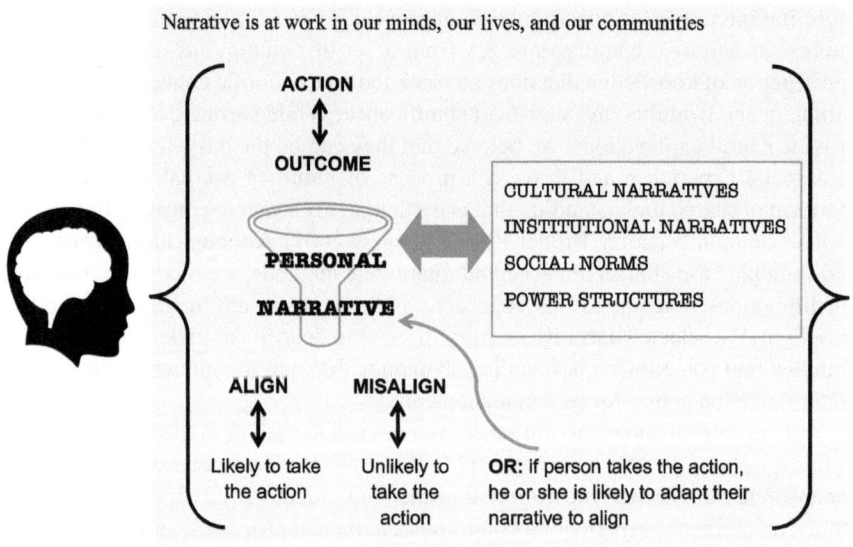

Narrative is at work in our minds, our lives, and our communities

Above is a simplified illustration of how narrative works in our minds. The double arrows represent the constant interplay between our minds, our experiences and the social narratives that influence us as we ever interpret and influence them. When interacting with people in community, people constantly share their personal narratives through their perceptions and through the actions they choose to do and not do. As two, or two hundred, or two thousand peoples' personal narratives collide, we can learn about each other, issues and ourselves, if we remain conscious of the role of narratives in our lives.

2.2 Why Is Narrative Important in Addressing Community Issues?

In this book, pest management is conceptualised as a community problem that requires collective action, which is achieved through development of a shared vision and a shared commitment to act. Framed this way, efforts to collectively address pest management issues can be seen as an exercise in community development. The goal of community development is to increase citizens' capacity to improve their quality of life (Green and Haines 2012). Practitioners in the community development field may focus on managing natural resources (Anderson et al. 2016), improving collaborative relationships between institutions and their surrounding communities (Bridger and Alter 2006), preserving and celebrating local history (Foth et al. 2008) or, as illustrated in this collection, managing pest species. Regardless of the issue under investigation, the power of narrative to increase understanding and empathy with others suggests, as Mayer (2014) puts it, that:

> Narrative is perhaps *the* essential human tool for collective action, a tool of enormous power and flexibility for constructing shared purposes, making participation in collective action an affirmation of personal identity, providing assurance that others will join us in the cause, and choreographing coordinated acts of meaning (p. 49).

As the collection of stories in this book illustrates, pest management is not solely a scientific or technical issue; it is also a social issue. Baiting, trapping and shooting are important management techniques, but evidence shows that they are weakened when collective action does not exist (Marshall et al. 2016; Ostrom 2010). For example, if only one landholder decides not to participate in a collective baiting program, the action of all other property owners involved in the effort can be fruitless. Narrative can foster collective action by highlighting shared values, illustrating possible futures and signalling assurance of individual commitments (Mayer 2014; Steffensmeier 2010). Stories connect people into groups who share expectations and who then can revise old stories and create new ones (Frank 2010). This book suggests that pest managers can work with diverse community stakeholders to build shared narratives in support of collective action. As readers, these stories invite your participation as well. Stories unite tellers and listeners, and offer opportunity for further conversation (Bradt 1997).

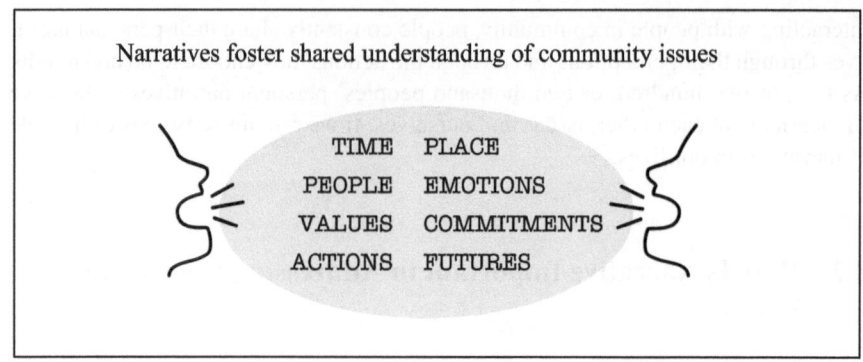

2.2.1 Framing the Issue Through Narrative

Framing refers to the process of developing a conceptualisation of an issue (Chong and Druckman 2007). Frames, or mental models or worldviews, reflect stakeholders' individual and collective philosophies and experiences. Just as photo frames set the borders of an image, mental frames provide a particular set of definitions and interpretations for viewing an issue. There is a dynamic relationship between narratives and frames. Narratives can create frames through which people view an issue; and frames can influence the narrative created about an issue. Framing theory shows that the way in which we approach an issue impacts our opinions and our perceived choices (Chong and Druckman 2007; Schmid 2004). For instance, if you view pest management through an animal rights frame, you might see shooting animals as unethical and be unable to support a narrative that rationalises lethal control. On the other hand, if you view the issue through an environmental rights frame, you may see how your passion for protecting animals overlaps with a narrative of protecting biodiversity through reducing species overabundance. By seeing the issue through both frames, you may then be able to consider the pros and cons of pest species management.

We suggest that narrative enquiry can be a useful tool for analysing current frames and developing new frames that embrace previously unconsidered ideas, navigate possible solutions to collective action issues and motivate and affirm support. Being alert to the way an issue is framed can help identify sites of conflict and of potential collaboration. Through awareness of competing or overlapping frames, we see the complex interaction between world views, subjective values and human complexities which can disrupt the scientific consensus and lead to surprising or unexpected pest management narratives. This ability to reframe an issue can be a positive asset for pest managers and community members who hope to motivate collective action amongst a diverse group of individuals. Understanding different ways of viewing the issue can give citizens the opportunity to experiment with others' frames and work together to create a new way of viewing the issue, thereby writing a new narrative about how people from across the community might be able to work together.

2.2.2 Community Narratives

Every community has a narrative. Through the telling, hearing and reading of stories, it is possible to gain a sense of familiarity with, and a common basis for talking about, particular places and their inhabitants. A community narrative creates and reflects a community's worldview and values, defines sets of possible actions and responses to issues and enforces power relationships by including some voices and excluding others. Understanding the historical, cultural, ideological, political and economic factors that have helped to shape community narratives over time can play a crucial role in understanding contemporary issues (Mayer 2014).

One unique kind of community narrative is the "heritage narrative". These historical accounts are especially important when it comes to creating a sense of common interest or collective identity (Bridger 1996). These are selective interpretations of the past that feed into and are partially driven by the demands, sentiments and interests of those in the present. Heritage narratives give temporal persistence to a community by providing an account of its origins, the character of its people (both past and present) and its trials and triumphs over time. The stories that are told about how a community came to assume its form provide an overarching framework, within which the meaning of contemporary events can be placed. The community, in this sense, "… is not different from the story that is told about it; it … is constituted by a story of the community, of what it is and what it is doing, which is told, acted out, and received in a kind of self-reflection narration" (Carr 1986, pp. 149–150).

While a community may report a dominant narrative around a common interest or sense of identity, this narrative may not be equally co-created or shared by all. Collective stories like heritage narratives reflect past power relations, while at the same time shaping and reshaping power relations in the present (Gaventa 1980).

The key question is: whose interests are most reflected in a dominant narrative and to what ends? People in positions of privilege shape the cultural and political narratives for better or for worse (McAdams 2008). The victors, the colonisers and the wealthy often write community, state or national histories, leaving little room for the stories of marginalised people (McAdams 2008; Urist 2015). While dominant narratives may shadow divergent accounts, community development practitioners must recognise and seek out alternatives.

Community narratives include, and are impacted by, political narratives. In turn, beliefs, values and emotions significantly impact political narratives (Swift and Dokecki 2013). Community practitioners can learn a great deal from exploring the inherently political dimensions of their work and the diversity of narratives within one community.

2.3 Why Use Narrative for Community Engagement in Pest Management?

Narrative enquiry is an appropriate way to explore human experience in the complex and changing context of pest management (Webster and Mertova 2007), because narratives reveal how people understand the world and act in it (Bruner 1990; Gergen 1999). Narratives can build shared understanding from which to address questions of community action. Narrative also provides a way for pest managers to understand the importance of context in facilitating collective action, just as meanings conveyed in stories are context-specific (Bradt 1997). In this collection, practitioners share stories of their adjustments moving from one project to another, showing how the unique community dynamics of each locale impact one's work. For instance, many practitioners detailed the importance of understanding a community's history and geography in order to better relate to residents. Practitioners told stories about understanding the political dynamics influencing a community, such as how diverse groups interact. Approaching farmers, animal rights activists, landowners, government officials and environmentalists, practitioners told stories about learning to understand the frames of others and to communicate across diverse interests. In this way, pest managers facilitated people to come together to develop shared frames (such as the desire to collectively manage pests), which inspired cooperative action, and which created new narratives about the community's capacity to work together.

2.3.1 Narrative Enquiry as a Research Methodology

Beyond the role of narrative in building the conception of self and community, narrative enquiry is also a well-established research methodology. Narrative enquiry can be thought of simply as "the study of experience as story" (Connelly and Clandinin 2006, p. 375), or as "an approach to the study of human lives conceived as a way of honouring lived experiences as a source of important knowledge and understanding" (Clandinin 2013, p. 17). In honouring the lived experiences of others, narrative enquiry can allow participants to gain agency over how their decisions and their communities are understood, as the research embeds opportunities for people to participate in co-creating and "re-authoring" the narratives that emerge through the research process (Du Plessis 2002). Narrative enquiry is fundamentally relational,

as it serves as "an exploration of the social, cultural, familial, linguistic, and institutional narratives within which individuals' experiences were, and are, constituted,
shaped, expressed, and enacted" (Clandinin 2013, p. 18).

2.3.2   Narrative and This Study

Narratives are also a research tool for gathering diverse perspectives that are not
always recognised or heard in expert-led processes of community engagement. Consistent with our belief in the value of socially constructed knowledge and respect for
diverse voices, we see narrative accounts as an entry point for not only collecting
stories but also engaging in a dialogue with the narrators. As we listen to stories,
we hear community members and individual practitioners share their insights from
years of close observation. These narratives are grounded in their particular social,
cultural, geographic and economic context and offer rich material for those who take
the time to read, question and reflect on what can be learned.

This book features both practitioner profiles and community case studies, which
were developed through narrative enquiry methodology, and which reflect the ideals
of narrative enquiry as an approach towards research. The researchers conducted in-
depth interviews with 52 land managers, agency staff and community members who
work to implement community-led pest management throughout Australia. From
these interviews, the researchers developed 12 individual practitioner profiles and
3 community case studies. The profiles and case studies explore how individuals
and communities perceive pest species and how they frame the issue of community
pest management. The profiles provide an in-depth look into practitioner motivation
and their perspectives on stakeholder relationships and land management. The case
studies build on the practitioner profiles to investigate how community groups engage
in the specific problem of wild dog management and how community members and
pest managers view their and other stakeholders' roles in it.

Practitioner Profiles
Practice stories (or practitioner profiles as we call them here) are useful learning tools.
They open a window into lived experience, illustrating challenge and opportunity, as
well as wisdom gained from this experience. As John Forester writes, "practice stories
can teach us about the complexities and uncertainties, the formal obligations and the
informal entanglements, the ambiguities … the strategies and sensitivities involved"
(Forester 2016, p. 339) in working at the interface between community interests,
scientific knowledge and pest management. To develop the practitioner collection,
individual interviews were conducted with 12 professionals involved in working with
community members to manage a range of pest species. These interviews were guided
by five open-ended questions that drew out their personal stories and positioned their
professional experience within this broader context. The resulting transcripts were
then edited, reshaped and re-storied in consultation with the practitioner, to develop a
first-person narrative that presented a cohesive story of their practice and experience

over the course of their career (Ollerenshaw and Creswell 2002). In these stories, "practitioners speak of what they did and learned in particular moments of their work and lives. They provide us with windows into the situational and contextual work and experiences of people who are engaged in civic life" (Peters et al. 2010). In essence, these profiles are stories of practice. They tell us about the broader relational aspects and challenges of community pest management, as well as the practitioners specific roles and actions within the context of their own lived experiences.

Wild Dog Management Group Case Studies

Similarly, the case study interviews use narrative techniques to explore how selected wild dog action groups were conceived, developed and sustained over time. The case studies reveal how these groups evolved to not only address wild dog management but other challenging social issues, such as community capacity, community power and resilience to changing institutional settings. A set of open questions and prompts encouraged 41 respondents to tell their personal story of wild dog management and their experience of the group response to this threat (Goodson et al. 2010). Each interview was transcribed and summarised in a template that identified key narrative moments of tension and resolution, key protagonists and events, and the respondents' portrayals of themselves within the broader context of the group and the wider community (Corvellec 2006). The summary documented instances where issues of wild dog management and community action were framed as either problems or opportunities, and ideas of "success" that were linked to these frames (Bold 2012).

A multiple case study design allowed case-by-case exploration and comparison of the different legal and policy settings; demographic, biophysical, sociopolitical and economic situations; and development of detailed community profiles (Johnson and Christensen 2008; Patton 2002). The case studies combined documentary sources such as policy documents, project documents, media sources and public-facing websites, with interview data to develop a data archive of each wild dog management group at a particular moment in time (Feagin et al. 1991; Rapley 2007). The research aimed to explore the lived experience of those engaged in wild dog management, generating rich qualitative data grounded in the specific circumstances and context of the subject (Holloway and Jefferson 2000; Flyvbjerg 2001). The research team members reviewed the results of the analysis with one another and with research participants in line with accepted standards for strengthening qualitative studies (Ross 2017). Detailed narratives and an accompanying analysis have been developed for each case study, along with an overarching comparative analysis.

2.3.3 Approaching the Practice Profiles and Case Studies

Within this chapter, we have conceptualised narrative as a way of making meaning from lived experiences, a practice and tool for community development, and a research methodology. This discussion has been guided by four critical questions about narrative: What is narrative and what are its origins? Why is narrative important? Why use narrative for community engagement in pest management? How is narrative specifically used in this study?

Narrative is how humans make sense of the world and their experiences. Through narrative, we assign meaning and identify relationships between ourselves and other people, organisations and larger narratives in our culture, society, community, geography and history. Narrative is critically important to our ability to live and work functionally as individuals and in groups, because narrative allows us to create a clear picture of our identity, to understand our place in society and to reconcile discrepancies between our internal experiences and external pressures. In community development, narratives can frame the way we think of an issue and inform our perceptions about what is possible. Narrative can support collective action by enabling individuals to understand diverse viewpoints, to develop shared interests, to decide upon a cooperative strategy and to tell a story that supports continued commitment. When pest management is viewed as a collective action problem embedded within layers of social context, it is possible to see the role of the dominant and alternative narratives of a community and how they impact management efforts. In this study, narrative offers a valuable lens through which to understand the complexities of pest management and to conceptualise potential paths forward.

Throughout the rest of the book, we hope you will become engrossed in these stories of practice and community. We hope that you will be transported into the shoes of the narrator—approaching a disgruntled community leader who holds a rifle, entering a room of animal rights activists to explain the ethics behind feral horse control or collecting bugs to learn about wildlife as an eager child. Continuing in the spirit of narrative enquiry, we encourage you to reflect on the relevance of these stories to your personal and professional experience. In your reading of these storied accounts, you may reach alternative conclusions to those which we present in our analysis. This possibility of alternative understandings is the dynamic essence of narrative enquiry, one which we embrace. The purpose of this chapter has been to remind the reader that each narrative is a construction and that authoritative versions can always be retold. We encourage you to explore the possibilities embedded in these stories.

References

Adler, J., & King, L. (2012). Living into the story: Agency and coherence in a longitudinal study of narrative identity development and mental health over the course of psychotherapy. *Journal of Personality and Social Psychology, 102*(2), 367–389.

Andersen, A. O., Bruun, T. B., Egay, K., Fenger, M., Klee, S., Pedersen, A. F., et al. (2016). Negotiating development narratives within large-scale oil palm projects on village lands in Sarawak Malaysia. *The Geographical Journal, 182*(4), 364–374.

Beck, J. (2015, August 10). Life stories. *The Atlantic.* https://www.theatlantic.com/health/archive/2015/08/life-stories-narrative-psychology-redemption-mental-health/400796/. Accessed December 12, 2017.

Bold, C. (2012). *Using narrative in research*. London, UK: Sage.

Bradt, K. M. (1997). *Story as a way of knowing*. Kansas City, MO: Sheed & Ward.

Bridger, J. C. (1996). Community imagery and the built environment. *The Sociological Quarterly, 37*(3), 353–372.

Bridger, J. C., & Alter, T. R. (2006). The engaged university, community development, and public scholarship. *Journal of Higher Education Outreach and Engagement, 11*(1), 163–178.

Brooks, P. (1984). *Reading for the plot*. New York, NY: Alfred A. Knopf.

Brown, J. (2005). *Storytelling in organizations: Why storytelling is transforming 21st century organizations and management*. Boston, USA: Elsevier Butterworth-Heinemann.

Bruner, J. (1990). *Acts of meaning*. Cambridge, MA: Harvard University Press.

Carr, D. (1986). *Time, narrative, and history*. Indianapolis, USA: Indiana University Press.

Chong, D., & Druckman, J. (2007). Framing theory. *Annual Review of Political Science, 10*(1), 103–126.

Clandinin, D. J. (2013). *Engaging in narrative enquiry*. New York, NY: Left Coast Press Inc.

Clark, W., Golder, M., & Golder, S. (2013). *Principles of comparative politics* (2nd ed.). Thousand Oaks, CA: Sage Publishing.

Connelly, F. M., & Clandinin, D. J. (2006). Narrative enquiry. In J. Green, G. Camilli, & P. Elmore (Eds.), *Handbook of complementary methods in education research* (pp. 477–487). Mahwah, NJ: Lawrence Erlbaum.

Corbett, E. P. J., & Eberly, R. A. (2000). *The elements of reasoning* (2nd ed.). Boston, USA: Allyn and Bacon.

Corvellec, H. (2006). Elements of narrative analysis. *GRI-rapport, 2006:6*. Gothenburg, Sweden: Gothenburg Research Institute.

Daiute, C. (2014). *Narrative inquiry*. Thousand Oaks, CA: SAGE Publications Inc.

Du Plessis, R. (2002). The narrative approach and community development: A practical illustration. *Africanus journal of development studies, 32*(2), 76–92.

Feagin, J., Orum, A., & Sjoberg, G. (Eds.). (1991). *A case for the case study*. Chapel Hill, NC: University of North.

Fisher, W. R. (1985). The narrative paradigm: An elaboration. *Communication monographs, 52*, 347–367.

Flyvbjerg, B. (2001). *Making social science matter: Why social inquiry fails and how it can succeed again*. Cambridge, UK: Cambridge University Press.

Forester, J. (2016). The challenge of transformative learning: mining practice stories to study collaboration and dispute resolution strategies. In R. D. Margerum & C. J. Robinson (Eds.), *The challenges of collaboration in environmental governance* (pp. 338–354). London, UK: Edward Elgar Publishing Limited.

Foth, M., Klaebe, H., & Hearn, G. (2008). New media and digital narratives in urban planning and community development. *Body, Space & Technology, 7*(2). http://people.brunel.ac.uk/bst/vol0702/home.html. Accessed December 14, 2017.

Frank, A. (2010). *Letting stories breathe: A socio-narratology*. Chicago, IL: The University of Chicago Press.

Gabriel, Y. (2000). *Storytelling in organizations: Facts, fictions, and fantasies*. New York, NY: Oxford University Press.

Garreau, J. (2017, January 3). Weaponized narrative is the new battlespace. *Defense One*. http://www.defenseone.com/ideas/2017/01/weaponized-narrative-new-battlespace/134284/. Accessed December 12, 2017.

Gaventa, J. (1980). *Power and powerlessness: Quiescence and rebellion in an Appalachian Valley*. Chicago, IL: University of Illinois Press.

Gergen, K. J. (1999). *An invitation to social construction*. London, UK: Sage.

Goodson, I., Biesta, G., Tedder, M., & Adair, N. (2010). *Narrative learning*. Oxford, UK: Routledge.

Gottschall, J. (2013). *The storytelling animal: How stories make us human*. New York, NY: First Mariner Books.

Green, G., & Haines, A. (2012). *Asset building & community development*. Thousand Oaks, CA: Sage Publications.

Hatfield, E., Cacioppo, J., & Rapson, R. (1993). Emotional contagion. *Current Directions in Psychological Science, 2*(3), 96–100.

Holloway, W., & Jefferson, T. (2000). *Doing qualitative research differently: free association, narrative and the interview method*. London, UK: Sage.

Johnson, B., & Christensen, L. (2008). *Educational Research: Quantitative, qualitative and mixed approaches*. Thousand Oaks, CA: Sage.

Leger, M. F. (2016). Exploring the bicycle metaphor as a vehicle for rich story development: A collective narrative practice project. *The International Journal of Narrative Therapy and Community Work, 2*, 17–25.

Leibovich, M. (2015, December 8). When the "narrative" becomes the story. *The New York Times Magazine*. https://www.nytimes.com/2015/12/13/magazine/when-the-narrative-becomes-the-story.html?_r=0. Accessed December 12, 2017.

MacIntyre, A. (1985). *After virtue: A study in moral theory*. Notre Dame, IN: University of Notre Dame Press.

Maines, D. R. (1993). Narrative's moment and sociology's phenomenon: Toward a narrative sociology. *Sociological Quarterly, 34*(1), 17–38.

Marshall, G. R., Coleman, M. J., Sindel, B. M., Reeve, I. J., & Berney, P. J. (2016). Collective action in invasive species control, and prospects for community-based governance: The case of serrated tussock *(Nassella trichotoma)* in New South Wales, Australia. *Land Use Policy, 56,* 100–111. https://doi.org/10.1016/j.landusepol.2016.04.028.

Mayer, F. (2014). *Narrative politics: Stories and collective action*. New York, NY: Oxford University Press.

McAdams, D. P. (2008). Personal narratives and the life story. In O. John, R. Robins, & L. Pervin (Eds.), *Handbook of personality theory and research* (3rd ed., pp. 242–264). New York, NY: Wilford Press.

Ollerenshaw, J. A., & Creswell, J. W. (2002). Narrative research: A comparison of two restoring data analysis approaches. *Qualitative Inquiry, 8*(3), 329–347.

Ostrom, E. (2010). A multi-scale approach to coping with climate change and other collective action problems. Leviathan: ZeitschriftFur Sozialwissenschaft, *39*(3), 447–458.

Patton, M. Q. (2002). *Qualitative research and evaluation methods*. Thousand Oaks, CA: Sage.

Peters, S., Alter, T. R., & Schwartzbach, N. (2010). *Democracy and higher education: traditions and stories of civic engagement*. East Lansing, MI: Michigan State University Press.

Polkinghorne, D. E. (1988). *Narrative knowing and the human sciences*. Albany, NY: State University of New York Press.

Rapley, T. (2007). *Doing conversation, discourse and document analysis*. London, UK: Sage.

Ricoeur, P. (1984). *Time and narrative* (Vol. 1). Chicago, IL: University of Chicago Press.

Ross, K. (2017). Making empowering choices: How methodology matters for empowering research participants. *Forum: Qualitative Social Research, 18*(3). https://doi.org/10.17169/fqs-18.3.2791.

Schmid, A. (2004). *Conflict and cooperation: Institutional and behavioral economics*. Malden, MA: Blackwell Publishing Ltd.

Smith, B. H. (1980). Narrative versions, narrative theories. *Critical Enquiry, 7*(1), 213–236. (Autumn).

Steffensmeier, T. (2010). Building a public square: An analysis of community narratives. *Community Development, 41*(2), 255–268.

Swift, D., & Dokecki, P. (2013). The construction of politico-religious narratives: steps toward intervention promoting human development and community. *Journal of Community Psychology, 41*(4), 446–462.

Urist, J. (2015, February 24). Who should decide how students learn about America's past? *The Atlantic*. http://www.theatlantic.com/education/archive/2015/02/who-should-decide-how-students-learn-about-americas-past/385928/. Accessed December 12, 2017.

Webster, L., & Mertova, P. (2007). *Using narrative inquiry as a research method*. London, UK: Routledge.

White, H. (1980). *The value of narrativity in the representation of reality*. Chicago, IL: The University of Chicago Press.

Part I
Practitioner Profiles: First-Person Practice Stories

Chapter 3
Profile Introduction and Analysis

Abstract Practice stories (or practitioner profiles as we call them here) are useful learning tools. They open a window into lived experience, illustrating challenge and opportunity, as well as wisdom gained from this experience. The profiles collected here contain useful first-hand knowledge of the intersection of pest management and community engagement in both personal and professional experience. While practitioners reveal diverse objectives, they share a common desire to bring people together, across their differences, for mutually beneficial and acceptable outcomes. The result is a collection that shows how the interaction between ideas, experience and reflection can sustain an individual in their work with community, over time.

3.1 Introduction

While many policymakers, researchers, land managers and landholders have a strong tendency to focus attention on biophysical and technical fixes for pest management, this book recognises the importance of community engagement, as a strategy for addressing pest management issues. These practitioner stories hope to address this tendency by increasing awareness of the benefits gained through a balance between the biophysical, technological and social dimensions of pest management. Our aim in this collection is to learn more about the human dimensions of pest management and, by listening to those who work "at the coalface", begin to describe and articulate a professional practice of community engagement in this field. The collection has relevance to policymakers, industry bodies, landholders and practitioners alike. All stakeholders involved in pest management, or indeed environmental management of any kind, can learn from these experiential accounts of community engagement practice.

Practice stories (or practitioner profiles as we call them here) are useful learning tools. They open a window into lived experience, illustrating challenge and

T. M. Howard et al., *Community Pest Management in Practice*,
https://doi.org/10.1007/978-981-13-2742-1_3

opportunity as well as wisdom gained from this experience. As John Forester writes, "practice stories can teach us about the complexities and uncertainties, the formal obligations and the informal entanglements, the ambiguities … the strategies and sensitivities involved" (Forester 2016, p. 339) in working at the interface between community interests, scientific knowledge and pest management.

The profiles collected here contain useful first-hand knowledge of the intersection of pest management and community engagement in both personal and professional experience. The collection and use of practitioner profiles to increase understanding of community engagement in pest management align with the narrative research approach detailed in Chap. 2 of this book.

The profiles were collected from participants in a collaboration between invasive species practitioners and university-based researchers which aimed to increase community-led collective action for vertebrate pest management. Peer recognition of practitioner expertise in pest management and community engagement was the primary selection criteria.[1] Many of the people interviewed share social and professional networks that support their work, and these influential linkages are evident in the text. All of the practitioners have either a natural science discipline or an agricultural enterprise background. While some have undertaken additional training or studies in community engagement or social science, the majority have intuitively built up a professional practice of community engagement through trial and error, rather than scaffolded through a recognised career path. Some practitioners participated in a short course at Pennsylvania State University that focused on developing leadership in community engagement. This exposed them to theories and concepts underlying different aspects of community engagement such as community development theory, diverse versions of community, democracy and the role of policy and legal settings as well as insight into invasive species management in the USA, where pest management is known as "wildlife management". During the course of the collaboration, some practitioners were also involved in designing, implementing and participating in research projects and formal learning networks that explored the relevance of community engagement scholarship to their work. From these relationships and experiences, practitioners gained new ideas and ways of thinking about the human dimensions of their work.

Readers will notice variations in the depth and detail of the collected profiles. This variation reflects the different communication styles of the individual practitioner and interviewers. Because the collection developed over several years, deeper

---

[1] The purposive sample was drawn from a long-standing network of researchers and practitioners that has been involved in the Invasive Animals Cooperative Research Centre (IACRC) program, which has a 17-year history. As a result, the sample has some demographic gaps—all practitioners were located on the eastern seaboard of Australia; there are only two women in the collection and no indigenous land managers. This represents an opportunity for future studies to capture these voices in this or other fields of environmental management.

relationships developed over time and this resulted in increased trust and familiarity between parties to the interview process.

3.2 Interpretation and Analysis

The process of developing, interpreting and making meaning from this collection has been a creative and reflexive one. As a research team, our interrogation of the practitioner profiles began at the interview stage. The original open-ended questions that shaped our enquiry were designed to encourage the practitioner to reflect on their personal and professional journey. The interviewer then asked additional questions throughout to probe more deeply on particular points and to seek clarification regarding important links between significant events and chronological details. In most cases, a second interview session was conducted, following up on details that may not have been fully explored in the first round. As the transcript developed, versions were shared with the practitioner to ensure that the story emerging through re-storying and editing of the transcript mapped to their lived and recounted experience. Additional points of clarification or expansion were requested, and practitioners provided final approval of the profile before publication. Our goal throughout this process was to maintain the voice of the individual practitioner. Although each profile has been edited for clarity in order to reduce repetition and craft a cohesive account, the text is directly drawn from the interview transcript.

Through a close reading and rereading of the profiles, the research team has focused on developing an overarching analysis, or meta-analysis, of this collection. We began to see that taken as a whole, the profiles are all concerned with the theme of change. Practitioners recount personal and professional experiences of challenge, growth and sometimes failure. They describe a desire to fundamentally shift the relationships between expert and non-expert knowledge, dynamics of community-led action, power and agency. We also identified a suite of practical lessons that support practitioners in this effort for change. These lessons are discussed in detail in Chap. 16, following the practitioner stories.

Overall, this collection speaks to the fundamental human experience of working with people. As we positioned questions in the context of the practitioner's life journey, it is possible to chart their development from youthful enthusiasts of the natural world to experienced agricultural professionals, research scientists and policymakers. Our interest and emphasis on the evolution of both their personal and professional journey enabled them to tell a story that connected their motivating passions with their professional work.

While practitioners reveal diverse objectives—some scientific, some related to agricultural productivity or biodiversity and others to community building—they share a common desire to bring people together, across their differences, for mutually beneficial and acceptable outcomes. The result is a collection that shows how the interaction between ideas, experience and reflection can sustain an individual in their work with community, over time.

The profiles demonstrate that achieving change is not easy—it requires a struggle that can sometimes leave the individual feeling isolated and unsure. We hope this collection provides inspiration and comfort for other change agents occupying this uncomfortable space between established and emerging ways of approaching community engagement for pest management.

Before moving on, we remind the reader that our interpretation of these practice profiles is not exhaustive; it is naturally selective, informed by our own worldviews and preoccupations. In the spirit of narrative enquiry, we acknowledge that there are other lessons to learn and new ways to understand the material. We issue an invitation to you, the reader, to investigate each profile for the rich insights that are relevant and meaningful to you.

We urge you now to accept the invitation issued by Darren Marshall[2] (see Chap. 12) and join us on this learning journey:

> My greatest challenge is, how do I engage a community on a topic that they do not rate high on their individual agendas? This is by no means only bringing the information and learnings I have to the table but also listening, learning and incorporating the community's knowledge and experiences into the mix. When everyone has had input and agrees on the outcome we wish to achieve as a community, the journey of feral animal management has begun.

Reference

Forester, J. (2016). The challenge of transformative learning: mining practice stories to study collaboration and dispute resolution strategies. In R. D. Margerum & C. J. Robinson (Eds.), *The challenges of collaboration in environmental governance* (pp. 338–354). London: Edward Elgar Publishing Limited.

---

[2]Senior Biodiversity and Pest Management Officer for the Queensland Murray Darling Committee (QMDC).

Chapter 4
Practitioner Profile (Lisa Adams): "We Cannot Carry the Whole on Our Own—We Have to Work Together"

Abstract Being a facilitator comes with the responsibility of finding value in all forms of knowledge that stakeholders bring to the table. Lisa Adams values different perspectives and takes them into account while addressing critical environmental and public health issues. Lisa views herself as a perpetual student, constantly seeking new information and adjusting her approaches. She came to appreciate the criticality of local knowledge while facilitating workshops in Southeast Asia. This was a formative experience for Lisa, even though the organisation she worked with did not integrate the community members' ideas. In her subsequent work on rabbit management in Victoria, she made sure that all stakeholders were heard, respected and included. She explains that her initial lack of knowledge about the rabbit management system was advantageous, because it encouraged her to seek out local knowledge. Adams' powerful dedication to democratic and participatory community engagement is the centre of her philosophy on problem-solving and innovation.

When I reflect on my personal journey, I see my life in chapters. The first chapter was where I grew up, on the coastal plain of Perth. We had a lot of personal freedom back then. I have memories of just being out having adventures at the beach and in the bush, and that laid the foundation of my connection to country. I was the oldest of four kids. My father was from a Sicilian family that emigrated to Perth just after the First World War, and he was the first person in the family to go to university. My mother's side of the family was fifth-generation working class Australian. She had to leave school when she was 15 because daughters were not educated beyond that age.

My mum became a Catholic in order to marry my dad, and we all got sent to the local Catholic primary school in the suburbs. We had a close connection to the church, and it was very influential in my early primary school years. It was the start of my interest in social justice—there were years of indoctrination about the starving millions in poverty. I was in a split class with my brother, who was dyslexic and had learning challenges. One time a teacher directed my brother to stand on a dirty patch of floor because that's all he was good for. The teacher then wanted my opinion on that in front of the class. I remained silent, and it made me angry. That was a strong moment for me, to recognise injustice, and the dissonance between what was preached and what was acted out in terms of Christian values.

My mother lost faith in the Catholic system because she saw that the system failed my brother. She sent me to a private Uniting Church school for a couple of years, which helped me understand what it feels like to be in the minority, when previously you'd been in the majority.

I tuned into different ways of thinking from a young age. What struck me most was the racism around me. Certain people would cause me grief if they knew I was half-Italian. There was a lot of racism, particularly towards Aboriginal people, and I felt the offense of that. My mum worked as a volunteer in the local boys' prison. I went to discos there, and the majority of the boys were Aboriginal. They were great guys. When I finally danced with an Aboriginal boy, he asked what institution I came from. Then he went back to his prison cell, and I went back to my white middle-class life. That was another eye-opener for me.

Eventually, my parents separated. They had the first cross-cultural marriage ever in this Sicilian family, and it was always going to be hard because of the cultural differences. I went to live with my dad in the city, and elected to go to the local public high school. The teachers were fabulous in both the private and public school systems. But what I loved about the public system was that everyone was treated the same. I appreciated being just one of the crew.

My mum partnered again and went to live down south in a place called Bow Bridge, where she ran a roadhouse 7 days a week, 12 hours a day. I would spend all my holidays down there as a teenager. The roadhouse was the post office, the liquor store, the petrol station, the only fast food outlet and grocery store for a 30 min drive north and a 40 min drive south—the community hub. We had sheep farmers, cattle farmers, fishermen, surfers and hippies—all of those characters would flow through our kitchen. It was a great celebration of humanity and community and diversity.

I had two career ideas after finishing high school: to be a journalist or to do veterinary medicine. My goal was to do something useful and worthwhile for society. I went for an interview to become a cadet journalist, and the guy who ran the interview sexually harassed me. It was a traumatic experience and my journalism ambitions ended at that moment. I went ahead and enrolled to become a vet.

Through vet school, I enjoyed the clinical role; working with animals and their owners, and the science was fascinating. I was also interested in the bigger social picture—food production, food systems and herd health. I studied social and political theory as electives, although I didn't really connect it to being a vet at first. The discipline that tapped that for me was epidemiology, because it's about understanding diseases and dynamics in populations.

Because I had all these interests, I juggled a lot of jobs after I graduated: I worked as a small animal vet in a suburban clinic and as a research assistant and tutor in the epidemiology unit of the University Department of Medicine at Royal Perth Hospital. I'd also go down south and do large animal work with the farmers near Bow Bridge. Being a vet is really just a licence to learn—you're always learning. I never had an issue with asking questions. In my experience, you've got to gain the trust of the farmer and, if they can see that you're genuine and you're working on their behalf, then they will work with you to find answers or solutions.

There were challenges. Once, I followed what was in the book but the guidelines weren't right for this particular animal. Within 24 h that dog was vomiting blood and the owner was really angry with me. I needed to hang onto my own knowledge and wisdom in that moment. Critical thinking is important, because when you get bitten by a wrong decision, you can learn from it.

My husband and I moved to Scotland for a couple of years, and lived in a village about 40 min south of Edinburgh. He was the local doctor and I was the local vet. I worked across the nearby villages and got a good dose of mixed animal practice. By then, we had our first baby, and that enabled us to get very close to the community in the village. If we could have stayed there we would have, but my husband was on a General Practitioner training scheme and when it ended there were limited job prospects, so we returned to Australia.

After our second baby was born, I went to work with my former boss, who by then was the Director of Epidemiology at the Health Department of Western Australia. I conducted epidemiological and statistical analyses which entailed detailed coding. It was really tedious and I wasn't passionate about it. At this time, the state government was starting to outsource a lot of activities—it was a period of transition. I could see there was a role for setting up research ventures that would help people collaborate across the system. I was young and bold, so I set up a consulting business to help pull together research involving the health department, and the local universities and research institutes. I relied on the content knowledge from my vet and epidemiology training, but the most useful skills were from facilitation training I completed during this time at a private college in Perth.

Then the perfect project came along. I was contracted by a consortium of research leaders to write a business case justifying industry and government investment in a proposed Australian Biosecurity Cooperative Research Centre (CRC) for Emerging Infectious Disease (EID). This work brought together my interests in public health, animal health and environmental health. The business case was successful in bringing more than $20,000,000 of new investment for research and education in emerging infectious diseases. I was appointed as the foundation Executive Director of the CRC. I recruited the Chair, an outstanding person called Mal Nairn. Together we appointed the Board, and set up all the legal agreements and the institutional rules about how this cooperative effort would happen.

Mal had previously served as Vice Chancellor for a couple of universities. I learned from him as a coach. I also learned enormously from Barney Glover, who was the Director of Research and Development at Curtin University at the time. I saw myself as an apprentice; I watched how they interacted, I watched how they thought, I watched their processes and I learned from them.

During this apprenticeship, I learned that governance comes down to processes and principles. How you begin something is absolutely critical, because that establishes the culture. Everything has to line up in terms of how you wish people to behave. You need to ask "What are the norms you want to establish?" That was very powerful training that I've held onto ever since. I was really privileged to learn first hand from these extraordinary people.

One legacy of the CRC that I'm really proud of was the support we gave to 60 Ph.D. scholars across Australia and Southeast Asia. Those scholars have become the emerging leaders of the region. This investment built the regional capacity for people to work in their own place, and also collectively and cooperatively across countries. When you invest in people and relationships, that investment keeps on giving.

After the CRC wound up, I went back to consulting work. I facilitated some research-planning workshops in Southeast Asia for an international development agency. There were a lot of scientists in those workshops who were not from the region. I could see how their expertise biased their understanding of what solutions were available to the local community. In the end, the solutions that were put forward through collaboration between the scientists and the international development agency did not reflect the local wisdom. I was shocked by this.

My facilitation skills were insufficient to carry that dilemma; I could facilitate the workshops, I could write the report, but the final outcome of the research investment did not make sense to me.

During this time, we moved to Melbourne and I was looking for a job to get to know the region. The job of National Rabbit Management Facilitator came up, and I thought, "I could do that". The job was funded by the Invasive Animals Cooperative Research Centre (IACRC) and hosted by the Victorian Government. A wonderful surprise was that right at the beginning of the job I got to spend 3 weeks at Penn State doing a short course about leadership and community engagement for invasive species management.

It was an absolute epiphany. Here, I was, in my mid-40s, making a very good living, but that good living was insufficient. I had questions and I had personal dilemmas around making sense of what I was seeing. At the course, I discovered scholarship relating to these issues and I was so excited that I was up all night reading.

The academics who ran the course were speaking the language of community development. I noticed the language; I noticed the way of engaging with problems and issues; I noticed how they were dealing with the complexity of the issues upfront, defining them as social problems rather than as technical problems. I watched how they communicated and how different types of knowledge came together. It just blew my mind and I came away deeply moved. Looking back on those EID-CRC years, we did good work, but we were falling short in terms of understanding the social and political dynamics. I think we were coming at it from a very narrow, scientific frame.

When I got back to Australia, I had a lot of decisions to make about the scale, framing and methods of the rabbit project. Because I was funded by the IACRC, I had the benefit of being part of a research program where questioning and experimenting were permitted and seen as normal. If I had been in a standard technical role, I would have been battling all the way. People would have asked, "What are you talking about? Why is this important?" But the Victorian State Government gave me a free rein. They wanted *community-led action on invasive species management*—that was their only instruction.

My ambition was to set up a participatory, democratic and system-strengthening approach to support community-led action for rabbit management in Victoria. System strengthening has been around for a long time in public health, but for some reason not in natural resource management. I already knew about participatory processes, because I'd been doing participatory work for 20 years. What was new for me was making an explicit reference to democratic practices. This was intentional, because once the word "democratic" is included, you've got to be really clear about the intent. You're thinking about power dynamics from the get-go. You're thinking about questions such as: Whose voices are heard? Who is excluded? How decisions are made and how information is shared?

The word "democratic" gets easily lost. Throughout the project, when other staff or stakeholders were talking about how the project was designed and what it hoped to achieve, the word "democratic" would just drop off, it would go missing. I would say, "We've got to put that word back in there. That word is important". People would ask, "Why? What does it mean? How does it affect the project?"

The way that I approached the project was driven by this idea of democratic practice. For example, in my experience of group work, everyone carries something of value. The person who is being very emotional is carrying the emotion on behalf of the group—that's a valuable asset. The critic is also a valued asset, because sceptical thinking is an important type of thinking. Each helps us test our knowledge and take better decisions. So, in the facilitation role I'm helping the group to understand that in one way or another. Conflict can be understood as bringing value to the group because we have to think more creatively about how we can work together, and how we can move through conflict to find solutions. In this way, conflict can be a source of innovation. As long as we bring respect for the other, it can work. As a facilitator, my role is to ensure that every individual in that group feels safe, and hold the group accountable to each other and the problem or task. I respect every individual in the group, and I assume that they bring valuable knowledge, insight and wisdom.

Because I was new to invasive species management, I began by asking "How does rabbit management in Victoria work?" No one could give me an answer. I thought, "How can we do anything if we don't understand how it works? We've got to understand how rabbit management works, we've got to speak to people". So I interviewed people from across the system. Farmers, pest controllers, scientists, local government–state government workers, environmentalists; and I learned there are many different groups involved in rabbit management.

There's a history. Rabbits were introduced into Victoria in the 1880s, and the State has been stumped by the rabbit problem ever since. There were stories about how people were driven off the land because of rabbits. They had to move to the city, the father became an alcoholic, there was abuse within the family—people were suffering the emotional and psychological trauma of the rabbit plague to this day. People in communities felt as though they didn't have agency over the rabbit problem. They got stressed, and that stress was acted out in all sorts of ways. They were angry and frustrated and they had become demoralised.

During this time, I did a week-long course on the practical and technical dimensions of how to control rabbits—how to blow up a warren, how to lay a bait trail and so on. The guys running the course were surprised that I wanted to do it, because they saw my role at the policy end of the problem, whereas I thought, "How could I even begin this project if I don't know the practicalities of how to deal with rabbits?"

At the course, there was a guy called Brad Spears who demonstrated how to explode a warren. As I watched him I had this sense that he knew a lot. So, being opportunistic, I asked if he would be willing to sit down with me for an interview. About 20 min into the interview he said, "You know Lisa, this is really cathartic. No one has ever asked me what I know". But it turned out that he knew more about how rabbit management worked than anyone, because he worked for all of the different players. He worked for National Parks, he worked for state government, he worked for farmers—he had incredible knowledge of how everything worked and how it all fitted together.

I interviewed around 23 people. I asked "What influences your decisions and actions? Where do you get your information from? Who do you trust?" I asked for permission to use particular extracts and compiled them into a 20-page briefing note which showed what the farmers were thinking, local government, state government, environmental consultants and so on, in their own words.

This is an example of how democratic practice has shaped this project. I believe in local knowledge. Understanding the wisdom of people in the communities who are dealing with the problem is important. The briefing note was compiled from interview extracts so that different types of knowledge were given equal space for consideration. If we are humble in what we know and what we don't know, we can reach out and learn from others. We cannot carry the whole on our own—we have to work together.

All of those people who had been interviewed were then invited to a workshop to try and figure out what we could do collectively. People were willing to get involved and were willing to donate their time and knowledge on the back of that very humble promise, because they were deeply motivated and cared about the rabbit issue. Everyone saw that it was an important problem from wherever they sat within the system. There was no favouring one knowledge over another; it was collaborative, not competitive.

It was a diverse bunch of people who had never met as a group before. I asked them upfront, "What did you think of the briefing paper? What did you learn?" The first comment from a member of the group was, "It's all about the philosophy. You've got to understand the philosophy of rabbit management". This person suggested that landscapes reflect how humans have acted in the landscape. And, that, as individuals and as citizens and as governments, humans have the power to change the landscape. They asked, "If we don't have a sense of how we want that landscape to be, how would we possibly know how to do rabbit management?" The group needed to agree on the vision for that landscape, and then think about what resources were available and who needed to be involved. That was an incredible beginning!

The group saw that the rabbit problem was way more complex than they had realised. Knowledge of the problem was incomplete, there were conflicting and competing interests in the landscape, and there was no simple solution. They might not have known what a wicked problem was from a technical point of view, but as a group they had just named it. So within 10 min the group had covered a lot of ground together. The briefing note had given everyone access to the same base level of knowledge and information. There was a shared understanding, which meant they could move straight to deliberation. They were able to move straight into thinking about rabbits as a complex system. They were able to move beyond their individual interests to think about the collective problem.

There was a power play early in the workshop. A community member named her understanding of the issue and a government person tried to shut her down by dismissing the comments as jargon and motherhood statements. They were suggesting that her understanding was incomplete or inaccurate. I said, "That's really interesting. What you've named is the problematic use of jargon and motherhood statements. I think we all need to own this idea. Let's write it up on the white board, because we all need to overcome jargon and we all need to understand the limitations of motherhood statements". The government worker paused, everyone paused, and then we got on with the business. We had dealt with the power play.

As a facilitator, I think deeply about what I'm likely to encounter in the group, what the concerns are likely to be and how to structure interactions. I use quite structured approaches. I draw on focus questions and I draw on tasks, because when someone has got a task to do, they can't keep going off on a tangent. I help the group to establish a goal. Without a goal, we're just flapping in the breeze.

So I asked the group, "If you were to intervene in the rabbit management system, what would you do?" About a dozen ideas came forward, and the group drilled down into those ideas. People worked singularly or together to draw pictures or systems maps of how they understood a particular idea, elaborating and making sense of it. It became apparent that they were all good ideas; there were no bad ideas. It surprises me that there is so much complex modelling or analysis around problems when often, if you put a bunch of people with diverse knowledge and relevant skills in a room and get them to talk through what they know and what they see, they come up with good ideas!

The next task was to take the learning from that workshop out to the wider rabbit management community to ask, "Do you think this is a good idea? Would it work? Do you see it differently?" This narrowed the dozen down to six ideas that were suitable for implementation. Then, I exercised my power in hand-picking a steering group with people from different parts of Victoria with diverse knowledge and skills and a good gender balance. This created a new citizen-led governance arrangement called the Victorian Rabbit Action Network. There was a bit of pushback from some people who felt that they should be on the group, but the Chairperson handled that very well. The group has been rolling out these six ideas and as far as I know nothing bad has happened—it's all good.

After a 2-year incubation period, the Rabbit Action Network has really started to resonate, particularly for government. The steering group has done some good work capturing and sharing knowledge about rabbit management across Victoria. The wider rabbit community has been activated, and government can see the benefits. The language is starting to stick. Now participants are starting to use the language of democratic process. The Victorian Government has received federal funding to extend this model across other invasive species. For the first time in my professional life, I'm hearing people say, "We need to think about how to embed principles of democracy in how we decide what happens to this money, and the governance around this money".

Invasive species managers need to be brave and experiment to find new ways of dealing with wicked problems. Skilled facilitation is important. Managers need the skills and confidence to bring people together, be confident that it will work and that no damage will be done. It's also about helping bureaucrats and scientists to see themselves as community members. Government likes to have a project plan upfront to justify funding the activities. The plan has to name outputs, deliverables and impacts, which can be difficult when you're working in a community context where the process is always evolving. When government projects are co-created with communities, they become more accountable than when a bureaucrat draws up a project plan or puts a line through a budget item.

I've learned that a lot of working with community is about "small p" politics—the politics of how we interact with people, how we listen, how people come to have empathy for each other and establish norms of reciprocity. In Australia, we don't really know what "small p" politics is, but it's something we need to learn. It involves faith that people can work together in communities to understand problems and work across their differences to come up with solutions. Particularly with volunteers, we have to show huge respect for the personal investment they make. That's an issue going forward—how do we manage the fact that people are volunteering their time alongside people who are paid? That's just an ongoing dilemma.

These are systematic issues that exist across our public sector agencies at the moment, but there are ways forward. We need to tap into the scholarship of democratic practice and experience to navigate those pathways, to innovate and change. It's going to take some leadership and collective action if we are to balance our investment portfolios along these lines. I feel as though I'm at a new beginning professionally. I think the next 10 years are going to be the best.

Chapter 5
Practitioner Profile (Ben Allen): "If People Don't Want to Do It, It's Not Going to Change Anything, so I Work with People"

Abstract Ben Allen combines expert knowledge with vital local knowledge in his work with communities. He sees that the "expert" status held by scientists—and scientists' detachment from the people impacted by their research—often undermines progress. Ben finds that collaborating with landholders is the most effective way to approach issues; farmers contribute valuable information about their community and the history that shaped it. Ben makes it a priority to meet landholders in their homes, even if it means driving 13 h. He seeks to conduct applied research that will be directly relevant and useful to the people with whom he is working. This reflects his belief that by working together, community members and scientists can address invasive species issues in much more effective ways than either can do on their own.

I wear several hats at the moment. I'm a Project Officer for Biosecurity Queensland, where I research peri-urban wild dogs or dingoes. My bread and butter comes from the Invasive Animal Cooperative Research Centre for this peri-urban dog project. We're investigating the ecology and management of dingoes and other wild dogs in peri-urban areas.

Another hat I wear is for the University of Queensland; I'm an Adjunct Lecturer there on dingoes, wild dogs, wildlife ecology and that sort of thing. I supervise several doctoral and other postgraduate students, studying aspects of dingoes and other threatened fauna. In my capacity as a university lecturer and researcher, I get to do things outside the scope of the peri-urban wild dog project, but related to it, like feral cat work or threatened species work.

Universities can often do things that governments aren't particularly interested in. I think in general—and this is a very broad generalisation—that governments have a very narrow set of interests, but when you're working on that interest you've got complete freedom over your time to do that work. Universities have a very broad set of interests, but you often don't have as much flexibility over your time because you've got teaching responsibilities or supervision responsibilities and you sit on committees and things. Having a foot in both camps means that I can be a lot more balanced. I don't get everything done personally, but I can at least supervise others.

© Springer Nature Singapore Pte Ltd. 2019
T. M. Howard et al., *Community Pest Management in Practice*,
https://doi.org/10.1007/978-981-13-2742-1_5

And so, for the benefit of all mankind we're doing better work than if I was just focused solely on one way of doing things.

Other than my work with the University and Invasive Animal Cooperative Research Centre, I'm also involved in other conservation and threatened species projects, so I consider myself to have quite a range of interests. I think it makes me a better ecologist and a better researcher, but it's tricky sometimes balancing all the different hats I wear. In the end, it all boils down to applied science; not just science for the sake of science, but actually really trying to resolve an issue.

One of the things that shaped my love of applied sciences was a report I read when I was a student called "The Seasonal Changes in Testicular Size of Male Captive-Bred Eastern Barred Bandicoots". Completely brilliant work, but absolutely useless for the real world, apart from maybe having some relevance for captive breeding programmes. I realised then that I did not want to do that kind of work. I wanted to do work that was useful for someone, which is why I like answering those applied science questions that reach across big, broad landscape scales.

I think another thing that shapes me is having a strong sense of working with land managers, like the farmers, or city council operators, or stakeholders at the bottom of the rung. It gives me a much better perspective than being inside an ivory tower, going into a national park where I don't have to deal with anyone, just doing my work and then going home. When I've got that applied science focus, I'm asking "How's this going to be good for the farmer?" or "How can I save a threatened species without imposing on the farmer?" I think it makes me a better researcher.

The grassroots people, the people who actually manage the land, like farmers, they are so critical to doing a good research job. They often have a lot of knowledge for their particular part of the world, knowledge that a scientist is never going to capture in 3 or 5 years of being there. You have to weigh up that knowledge with a scientific hat on, but they really do have so much knowledge on context, which you just can't get from a short-term experiment. So, they are a good ally and a good source of background information and support. I can't always do all of the activities I need to do by myself; I need landholders to work with, so I try to develop really good one-on-one relationships. In my experience, I can't do that over the phone. I've just got to have those face-to-face meetings.

Sometimes I feel that, if only there were no people, I could get a lot of things done! Although a lot of my work is autonomous, most of what I want to achieve for the benefit of the world, like saving animals, is hindered by people. But I can't save the world by myself. I have to work with people, to harness people. A lot of what science does is generate goodwill, motivate people to action and harness more troops. I'm interested in good quality science, and the rest of the world should be too, because that's what's going to save it. At the same time, I might have the best discovery in the world, but if people don't want to do it, it's not going to change anything. So, I work with people.

I'll never forget one time when I presented a public workshop on dingo issues to try and solicit participation from some big cattle station properties that we could work on. There were a bunch of property owners at the workshop and one came up to me and said, "I'd love that work to be done on my place" and I said, "Great, I'll come and do that". She said, "Do you know where I live? It's 13 h' drive away from your office, and no one from any agency ever comes up here". I said, "That's fine. *I'll* come up". They were shocked that any agency staff member thought that they were worth driving 13 h for. That experience made me think, "Man, researchers are just missing out on some really good opportunities" because they think, "Oh, it's too far" or "It's too hard" or "Oh, it's too hot" or whatever. I didn't want to be one of those people.

I also grew up with all this. My father is a zoologist, so from the time I was 6 years old, I was doing dingo research. I'm one of the few 35-year olds who has got 29 years' experience in dingo research! I went to university and got a couple of degrees, and then started doing it myself. During my undergraduate degree, I was also working for the university, tutoring and volunteering at the Queensland Museum, and collecting skulls (which I do as a hobby). From 2006, I started doing full-time research and then straight out of university I worked for the sugar cane industry on invertebrate pests—mainly white grubs, but also pigs and other small rodents. Then in 2008, I went back to dingoes, because that's what I like. That and handling animals. I have what I call "catch-it disease", so if it moves, I just have to catch it. It doesn't matter how dangerous it is or how stinky it is, if it moves, I just have to hold it and catch it. I loved catching cane toads and reptiles as a kid, it was great fun. I would put them in my pocket and take them to school.

Given this upbringing, I think I would drive myself nuts if I was working in just a small patch of bush for a city council. I like working on big, broad-scale problems in places where there's a proper horizon and a sunset. I don't get that in the cities. I like driving off, 5 h from the nearest person, and sleeping out on the ground in a swag, without a tent, for 2 weeks all by myself. I've got to do that to work on these big problems; I can't do it from a desk with crowd-sourced data. If I want the good stuff, I've got to go and put in the hard yards to get it across a whopping big country. As a scientist, I like the challenge of resolving a big, broad-scale problem, and dogs are a big, broad-scale problem. We're talking about populations across hundreds of thousands of square kilometres impacting big industries, like the cattle and sheep industries, which are worth billions of dollars in exports. We've got threatened species across all of that terrain, and dogs are part of that problem; they might even be part of the solution.

For the issue of wild dogs, which is so polarised, I get flack from both sides regularly, and both sides can be just as bad. There are really extreme farmers who think they've got to "Kill, kill, kill". They're just as wrong as the people who say "Save, save, save". It's about finding the right solution for the problem in context, and both sides have got to be prepared to bend a little. That might mean someone has to do something they don't want to do. The role of science should be to ask, "Well, how do we provide the right information to help people make those decisions?" Often, as a scientist, I'm in the middle, because if I say something pro-dingo, all of the anti-dingo people hate me; if I say something anti-dingo, all of the pro-dingo people hate me. I figure if both sides hate me, I'm probably getting it right. I'm just trying to do a good job.

I've worked for some terrible organisations in terms of organisational structure and stifling creativity and progress. Working for those stifling agencies is just horrible. Personally, it drives me bonkers and really gets me down—depressed and unmotivated and all those sort of things. But when I work for an agency that is a "yes" agency, then I become so much more productive and do a better job. I've worked for some fantastic organisations that really gave me intellectual freedom and allowed me to progress some of those things. They just have an attitude of, "Well we have this barrier, so how can we get around that?"

The motivation behind what I do at the moment for the Invasive Animal Cooperative Research Centre is getting the ecological data necessary to make evidence-based management improvements that help end users fix problems in a much more cost-effective way. I've got to juggle a lot of personal relationships at all sorts of levels to get that on-ground work done.

Over time you develop, you get better at it. I'm quite good now at handling media, particularly negative media, because it's really easy just to default to my research. I might have my own personal views, but when I'm pushed into a tricky corner it's best to come back to my own science. Just the wrong word could get me sacked. So, if I'm in the media only talking about my research, I feel like I'm on safe ground. I try not to engage too much more beyond what I can support from my own work.

What hacks me off is when people who have got no idea what they're talking about get put up on pedestals as wildlife experts, when they've never got out of the office in their life. They get interviewed as experts, put on committees that decide who gets what funding, put on decision panels for how a species is managed or put on scientific advisory groups for governments. It's the people who have never done it that are deciding whether a recommendation is good or not.

As an ecologist, the danger I see in this is that they'll potentially go down the wrong path completely, species will go extinct and farmers will go broke, all because of a viewpoint of just a few influential people. I'm not personally affected—it's not that I'm trying to compete with them at all—but I do want to make sure that people who know what they're doing are included in those decision-making roles. So I've got to get out there, be in the media, use social media, be "famous", to get a spot on that stage.

Online, I try and communicate only through *The Conversation*. It's a news media site that uses content sourced from academics and researchers, and allows other people to make comments. It's a great way to channel discussion or be in charge of the discussion. Social media is completely out of your control, but on *The Conversation* you're a little bit more in control of how the discussion flows—you can keep people on topic and you've got more space to explain.

I'll use it as part of a communication strategy for an article, so I usually only post when I've got an article to back it up. We'll coordinate with the publisher when the article is coming out. Then, the media units from the University of Queensland, Invasive Animal Cooperative Research Centre or Department of Agriculture and Fisheries will be ready with their key messages. I will have set aside time that week to not be in the field so I can respond to the online discussion. The print media can just lift content from *The Conversation,* because it's a creative commons licence, so they don't need any more quotes from me. I see it as a really good strategy, and people get right into the discussion on *The Conversation*.

It's a challenge to detach emotion from science; I've got to be passionate about what I do, but I've also got to be objective in what I'm looking at. Generally, if someone makes a comment on *The Conversation* that contradicts what I think, I'd like to think I'm big enough to change what I think rather than just disagree. I'm not trying to convince someone against their will. What I am trying to do is to make sure decision-makers have more than dodgy science, by people who have never left their office, to go on. Provided my work has been published, it's there to be considered—no one can say there's consensus, and they're less likely to adopt a law or change the law if there is no consensus. So, I have a job to do in bringing balance to the decision-making process, to stop decision-makers going down the wrong path or at least slow it down.

I'd like people to be evidence based in their decision-making, because that's what we need to fix the problems. We've all got to live in the world, we've all got to share it. We're in a situation where most of the country is ecologically challenged, so we have to make the right choices to turn it into something else or keep it the way it is. It's just about finding the right balance and not being too absolute in everything, for the benefit of everyone. You can't be out there saying, "The government is useless". It's not constructive. You've got to learn to play the game, with a bit of a thick skin.

I'm not too fixated on trying to rid the continent of invasive species. I don't want any new ones, and if there are ones that have just recently become a problem, well then we should do what we can to get rid of them. But once the horse has bolted, I'm not particularly fussed; it then becomes about what we can do to actually fix what we've got.

Ideally, I'd change the law to make dingo control voluntary. Some people aren't going to like that idea, but that's what I'd do. I'd still be just as active in dingo control in the places where dingo control is needed, but policy should allow people not to be active if they don't want to be. I'm an advocate for land managers to be their own type of scientist. One of the things we always said to the sugar cane farmers was, "If you think that variety of sugar cane works best, why don't you plant half a field of that variety and half a field of your other variety and see which one works best?" That's a scientific experiment done by the farmer.

For cattle and dogs, I would say to the farmer, "On that half of your property, manage them this way, and on that half of your property, do the other. Record what your branding rates are and do your own experiments". I'm a real advocate for those at the grassroots levels doing it for themselves. They don't need a government or researchers to tell them. If they're doing science for themselves and they're doing it well, they'll get their own solutions.

Chapter 6
Practitioner Profile (Dave Berman): "Building Trust with Community Members"

Abstract What drives engagement practitioners to persevere in the face of intractable issues? In many cases, it is passion. Dave Berman exemplifies a passionate practitioner. He believes that good science is essential to good community engagement. This unwavering commitment propels him through difficult situations in his work on feral horses, rabbits and cats. He first learnt about feral horses as a young man, developing a long-lasting affection for them. This led him to research their impacts and use this science to liaise between horse conservationists and pest management organisations, two groups that often have opposing goals. Although it is not always easy, Dave finds creative ways to communicate scientific knowledge to landholders and community members so they can find common ground. Dave views this as useful, applied work. Ultimately, Dave defines success as bringing science to the table so communities can co-create goals and solutions that make a real impact.

My family lived in Armidale, New South Wales (NSW) until I was eight, and then we moved to Wellington, NSW, where I lived until I completed high school. Dad was in charge of the Soil Conservation Research Station there. He had been a soil conservation officer in Armidale, where he built contour banks, a way of sculpting the hillsides to alleviate soil erosion. He was one of the first people, I think, to use that technique. He was a greenie—an environmentalist—before greenies were invented.

Dad had been riding horses since he was a young kid. He grew up on the New England Tablelands, a high-altitude district in northern NSW, where he used horses for stock work and as a contract plougher with draft horses. He joined the Light Horse Brigade, a mounted division of the Army, before World War II. "If I had to go to war", he said, "I wanted to go on a horse". This passion for horses was in my younger sister and me too. I can remember putting pocket money away, saving for a horse when I was 6 or 7 years old. When I was about 12, dad bought us a little white pony. At first, we didn't have a saddle. Dad said, "I'll show you how to get on". He vaulted on bareback and the pony bucked him off. So he vaulted on again and, about the third time, he rode the buck and stayed on the pony. That was my first lesson in horsemanship. "Get straight back on when you fall off." We got a saddle eventually and I became addicted to horse riding and in particular jumping horses over fences.

© Springer Nature Singapore Pte Ltd. 2019
T. M. Howard et al., *Community Pest Management in Practice*,
https://doi.org/10.1007/978-981-13-2742-1_6

In my fourth year of high school, my father asked me, "What are you going to do when you leave school?" I didn't really know the answer. Dad suggested I should be a biologist like TV star Harry Butler, an Australian wildlife scientist. I thought, "Yeah, that sounds good". So, I started working on prerequisites: maths and science, and suddenly I was top of the class in those subjects, which had never happened before. It's amazing what a bit of study can do. I went to the University of New England (UNE) in Armidale, and completed a science degree with Honours. I started thinking of doing a Ph.D., because I really liked finding out new things.

I wanted to be a research scientist, and I thought it'd be great to combine my passion for horses with research. I wanted to understand the natural behaviour of horses in the bush. I thought this would help me better train my show jumpers. One of my first showjumping horses had been a wild horse, an Australian Brumby. He was exceptional. There was something special about him perhaps because he grew up free in the bush. That experience got me interested in the behaviour of Brumbies and studying wild horses became my goal. I was always trying to build on the past, brick by brick, combining my experiences to build a stronger base of knowledge and skills.

So, I approached Peter Jarman at UNE and asked if he would be my supervisor and I applied for a scholarship to do a Ph.D. on Brumbies. I didn't get a scholarship at that time, so instead I did a Diploma of Education. Then, I taught for a year at Duval High School in Armidale. At the high school, I was a science teacher, maths teacher and sports master, among other things. I come from a long line of teachers: my mum and sisters are teachers and my grandfather and great aunt were teachers too. I think that's in me. I get great pleasure out of sharing what I've learnt from hard-earned experience.

By chance, one of the young fellows I taught was Simon Jarman, Peter Jarman's son. One day Simon brought a note to school from his dad asking, "Would wild horses drag you away from teaching?" Peter had secured money and a contract in the Northern Territory (NT) to look into the economics and ecology of feral horses in Central Australia.

As a result, I reapplied to do a Ph.D. with Peter as one of my supervisors. I moved to Alice Springs, a town in Central Australia, and began to look into a number of questions, including: "Where are the brumbies? How many are there? How much impact are they having on cattle?"

We were part of a larger research program Peter designed in partnership with Ken Johnson at the NT Conservation Commission (NTCC). Ken was an amazing public service boss who could achieve so much in a calm but determined way. The Conservation Commission was a funny name for the NT national parks agency. The agency combined conservation objectives with pest control, which is the way it should be. I was lucky to work with the unique members of the Wildlife Research team in Alice Springs. For example, Peter Latz, a botanist who grew up with Aboriginal people at Hermannsburg. Peter was a very practical scientist with an exceptional understanding of the ecology of Central Australia. Peter and the group generally valued research that resulted in practical outcomes rather than just scientific publications.

The initial goal of the feral horse research was to use aerial surveys to estimate the density and distribution of Brumbies in the NT. During the surveys, we covered most of Central Australia, from Tennant Creek and south to the South Australian border. On my first day of the survey, we flew for 10 h at 250 ft above the ground in a fixed-wing aircraft counting horses. I used the aerial survey data to help select a key study area, The Garden Station, to look at Brumbies in more detail.

My experience with this research has had a massive influence on the rest of my career—it was probably about three or four Ph.D.'s by today's standards. I looked at the social organisation of horses and their population dynamics. I looked at the dietary overlap with cattle, and the interaction between diet and habitat. On top of all that we included a study of the environmental impact of horses which we realised, as the project progressed, was essential for justifying a reduction in horse numbers. Then, I brought all that together to ask, "How can we manage these beautiful but overabundant animals?"

The owner of the property where I did most of my Ph.D. studies, Jim Turner, had a big influence on my research. He used to say, "What's the use of your Ph.D.?" That made me determined to apply the work I was doing. I wasn't just going to do a science project and publish it; I had a really strong drive to make a difference and—as he put it—be useful. I've had that same drive for every other project I've worked on since. I actually haven't published my horse research in peer-reviewed journals (yet), because my main aim was to achieve practical outcomes. I think the reputation of scientists suffers from a lack of practical outcomes. Ideally, we need both publications and practical outcomes but it is difficult to do both satisfactorily within the time required. To maintain the momentum of a management program involving community members, you often cannot wait for the scientific review process. Also, there is often valuable information useful for management that doesn't fit the requirements of a scientific publication.

Jim Turner gave me a hard time for at least 18 months until one day I turned up to conduct horseback transects through the rugged part of the property and I came across the Turners mustering cattle into a yard. I asked if I could help them. They said I could but they were nearly finished. So, I jumped on one of my horses, bareback (no saddle) and rode out to help. Although they were close to the yard and almost had the cattle in the gate, the cattle had other ideas. The cattle split up instead of going into the yard and went in all directions. I galloped after one of the steers, jumping gullies and crashing through the scrub. I managed to shoulder the steer to a standstill and helped bring him back to the yard. At last, I had demonstrated that I was useful and that I appreciated the difficulty of their job. We gained mutual respect that day and I was trusted at last. In fact, from then on I was treated as one of the family.

Before I started the research, I had not imagined that it would involve working with people so much. I thought I was going to do a Ph.D. and simply learn about horses in Central Australia. The NT Conservation Commission and the pastoralists wanted to go in and shoot these horses to get the numbers down to protect the environment, but they knew from what had happened in North America that they couldn't just do that because there would be community opposition. They had to get the essential ecological information to get the animal welfare groups and the horse lovers onside. Getting those groups onside increased the possibility that they could meaningfully reduce horse numbers using previously unpopular methods, like shooting from a helicopter or transporting to abattoir. We used science to show the truth of the matter. When representatives of those welfare groups came out to Central Australia, we showed them the carcasses of horses that had died of thirst and starvation. In presentations, I used hundreds of slides, instead of words or graphs, to communicate the science. I showed pictures of 80 carcasses around a waterhole in the Nineteen Mile Valley north of Kings Canyon, and said, "This is what happens if you do nothing". Showing those photos had a big impact.

By engaging with the interest groups, we gained agreement on a common goal. The common goal was to reduce the number of horses. For the horse protection groups and animal welfare groups the reason was to prevent overabundant horses starving to death. For the cattlemen, we were helping to increase cattle production and for the environmentalists we were protecting the native plants and animals and the soil. With this support, we reduced Brumby numbers by 70–80% across all of Central Australia during the time I was there. In Finke Gorge National Park, we were able to remove all the horses and many of the surrounding properties were free from the impact of feral horses when I left Central Australia. Most of the remaining 20–30% of Brumbies were on Aboriginal land where today the problems of overabundance remain.

In hindsight, I realised the research program, initially developed by Peter Jarman and Ken Johnson, then refined as it went, followed a seven-step model for pest animal management, which I continue to use today. The model is similar to that documented in the book that we developed from the Central Australian program, *Managing Vertebrate Pests: Feral Horses* (Dobbie et al. 1993). The first step is gathering basic ecological information about the distribution and abundance of the pest. Step two is to determine the negative and positive impact of the pest; this is an important step that many people bypass. In a sense, you are defining your problem in this step. Step three is assessing the different control methods. Step four is consulting with all the relevant interest groups. Step five is implementing the agreed upon actions. Step six is evaluating your progress. Then, step seven is repeating all steps until you reach an acceptable level of species impact. These steps can overlap with each other, and they do not necessarily occur in this order; however, step four and five cannot be done well without some progress towards completing steps one to three. All steps are required and none are more important than the others.

Peter Jarman has a special understanding of ecology and pest management. Not only was he an exceptional scientist and teacher, he could also clearly explain ecological concepts to people. He was amazing at explaining how the plants and animals were interacting and what was shaping the movement and behaviour patterns of the feral horses. Some scientists get stuck on all the fancy statistics. They shove data into a black box, give it to a statistician, and then they get the answer out, not really understanding what happened to their data. Peter taught me basic methods of analysis, and the steps he went through were so logical and simple, I carry that on in my work today.

Science is important to find the facts but then these facts need to be disseminated to the various groups in appropriate ways. My family was often involved in musical comedies or plays. My mother played the piano, dad the violin and harmonica and my sisters acted and sang. Dad recited poetry and trained me to do so too. "Speak up son, there's no use reciting poetry the audience can't hear!" he would say, loudly. These skills have been very useful for presenting science to all sorts of audiences. In fact, my first scientific conference presentation at the 7th World Congress on Animal Plant and Microbial Toxins was in the form of a poem called "The Venom of the Sea Snake". I had no slides or data, just a poem summarising the work. I like to entertain as I present my science but it is vitally important that the science remains as solid as possible and is strongly supported by data.

During my time in Alice Springs, I met Will Dobbie. Will worked as a Range Management Officer with the Department of Primary Production. He is a scientist who can write songs and sing. When I started working for the NTCC we contracted Will to continue the Brumby work. We came up with some creative methods to communicate our science. In 1990, we were in a documentary on the Brumbies, called *Brumby: Horse Run Wild*. It's still available on YouTube (SkyVisuals 2013). We thought we were famous scientists in the documentary but I think we just made people laugh. A newspaper review said that we looked like we were out of a Monty Python movie! The documentary was shown internationally. In the documentary, we were running around in the bush with radio tracking gear, darting and collaring horses. It helped generate publicity and helped interest groups understand the problem. It shocked people to see horses dying of starvation and being shot from helicopter but the message was balanced and the science was illustrated through images and music that communicated our concern for the welfare of the Brumbies. The documentary painted a picture that people could relate to.

It's important to have a very good reason for doing what you're doing and let all interest groups know that reason and hopefully then they will accept that something must be done. For example, the Guy Fawkes River National Park (NSW) shoot in 2000 was a case in which the New South Wales Parks and Wildlife Services ("Parks") culled over 600 Brumbies in an aerial shoot. There was no feed for the horses because a bushfire had burnt through the Park. There were too many horses and they were at risk of starving to death. Although the shoot was well planned, Parks skipped over the consultation step. They wanted to take action quickly. They hadn't consulted sufficiently with neighbouring landholders who opposed the management action. They hadn't consulted with the animal welfare groups sufficiently. As a result, the aerial cull came as a nasty shock to many interested people and there was a lot of controversy about the management action. Helicopter shooting of Brumbies was banned in NSW and Queensland, which was a terrible outcome in terms of Brumby management.

A number of horse protection groups grew out of this event and these groups are still active today. I have been working closely with one such group, the Southeast Queensland Brumby Association, catching and rehoming Brumbies since 2009. Unfortunately, this method alone cannot solve the problem, and the horse numbers are still growing. There are probably more horses in Guy Fawkes National Park now than before the shoot. The horse protection groups fight against every control method except catching and relocating the horses or fertility control. But you can't succeed in most parts of Australia with these methods alone.

The humane aspect of Brumby management is significant to me on a personal level; I love horses, some of them have been my best friends. A solution to the Brumby problem will not be found until the impact of the Brumbies is thoroughly and scientifically documented. This science needs to be conducted with the horse protection and environmental groups working together with the scientists.

While I was in Alice Springs, I got married. My wife had some kids already (Kristi, Sharnee and Luke), and we had a couple of kids together (Carli and Jodie). The girls all rode horses in Alice Springs and Luke played Rugby League. To compete in showjumping, we used to drive from Alice Springs to Darwin and back. That's a 3000 km—a bit under 2000 miles—round trip. We also drove down to Wagga Wagga in NSW, where Sharnee represented the NT against the other states. We drove down there and back (over 5000 km) in a very slow old truck that couldn't go faster than 80 km—about 50 miles—per hour. It became apparent that if we were to compete in showjumping we needed to move closer to the shows.

In 1997, I applied for a job with the Robert Wicks Pest Animal Research Centre in Toowoomba. I had wanted to work there since I first heard about it at the Vertebrate Pest Conference at Coolangatta in 1987. I remember thinking at the time, "That's the place where I should be, because that's the kind of work I do". The Centre specialised in invasive animal research. John Robertshaw, a respected scientist, had just left and I took over his role doing rabbit research. Being able to more easily compete in showjumping was a big reason for our family to make the move from Central Australia back to the eastern states. We were able to travel to over 30 shows per year and the daughters became very competitive, winning Queensland, NSW and Australian Championships. I also managed to win a couple of Grands Prix and completed a World Cup Qualifier on a Brumby from Central Australia called Hawkeye.

When I started in the new job, I wanted to show everyone how I worked. I presented my plan to the Rabbit Advisory Group, which was comprised of the bosses in our department and people interested in rabbits. I had a diagram with a box for each group involved in pest management—scientists, landholders, operational staff and so on—and I showed that to succeed in managing rabbits you need all the parts of our community to work together, and that none are more important than others. "This is how I work," I told them, "The science is just part of it, and all these other parts—the landholders, the media, the operational staff—all have to work together to make a difference". This diagram was a simple systems map.

One mistake I made was to leave out a box for "policymakers" because in the Territory they hadn't seemed very important to getting good on-ground outcomes. But in Queensland, not considering the policymakers sufficiently resulted in future barriers to success and difficulty collaborating with sections of the department. Looking back, I realise I should have included policymakers more in my project engagement strategy.

In the rabbit research role, I was able to put all the learning from my previous work into practice. Following the seven steps of best practice pest management, I first mapped the distribution and abundance of rabbits in Queensland. The map clearly highlighted the key areas to target. One of the areas was a property called Bulloo Downs. According to the map, this one property, Bulloo Downs, had 25% of Queensland's rabbits; this property was possibly a source of rabbits for the whole southwest Queensland region. My team and I used science to help the landholders understand that rabbits *were* causing considerable damage to cattle production and that they *could* be controlled. Stanbroke, the pastoral company that owned the station, provided half of the funding; the government provided the other half, and all together 58,000 rabbit warrens were destroyed. We managed to take a place that had 25% of Queensland's rabbits down to only 0.25% of Queensland's rabbits. This was possible because we targeted the 4% of warrens on the property that the rabbits required to survive severe drought conditions. All warrens within 1 km of permanent water were destroyed.

The benefit to cattle production was estimated to be over $5,000,000 up to 7 years after ripping. Such a success would have been impossible without the hard work of so many people; the cooperation of the property manager, Geoff Murell, and support of the owners. My Robert Wicks Pest Animal Research colleagues put in incredible efforts under sometimes very unpleasant, hot, cold, wet or dusty conditions. Mike Brennan helped with the research and spent many days spotting for the dozer driver. Peter Elsworth provided logistic support for the research expedition teams involving up to 12 members for 10-day stints. Our boss, Joe Scanlan, allowed it all to happen and provided advice on vegetation assessment and pasture production and utilisation, and rabbit population modelling.

Working with the Robert Wicks Pest Animal Research team was a great experience that came to an end when I took a voluntary redundancy in 2013. After almost 30 years demonstrating the importance of science for pest management I was very disappointed that the newly elected state Government decided to reduce our research team and funds by 40%. The Robert Wicks Pest Animal Research Station was closed. Something that had taken decades to establish was partly dismantled within a few months. Fortunately, some very good people remained in the Pest Research team and they are building back towards the critical number of staff required.

After this, I took up the position of Regional Pest Technical Officer with the Queensland Murray Darling Committee (QMDC), which is a regional Natural Resource Management (NRM) body for the state. In this role, I drew on all my experience working with invasive species to help improve management of feral pigs, wild dogs, foxes, rabbits, feral cats and Indian Myna. We assisted the small community at Inglewood to reduce the Indian Myna population to very low numbers and we were able to measure an increase in the abundance of native birds resulting from the Indian Myna control. We devised a monitoring system to measure the benefits of wild dog and feral pig control programs using camera traps and I developed an app to quickly look through and quantify the hundreds of thousands of photographs taken. West of Moree in NSW we showed that feral pigs reinvaded areas within weeks of a helicopter shoot, whereas around Moonie in Queensland two consecutive shoots reduced the feral pig activity by 88%. We showed that a combination of baiting and trapping suppressed wild dog numbers by up to 83% on gas company land north of Roma. We demonstrated that effective monitoring could be conducted without compromising the control activities. Some people prefer to conduct pest control without measuring their performance and pretend they are doing a good job. Perhaps they are but no one knows unless you monitor.

In an attempt to find a method to beat perhaps the most difficult pest animal in Australia, the feral cat, I acquired my best friend yet, "Sophie" the Springer Spaniel. One day at a field day we were discussing how best to control feral cats, and Tom Garrett (QMDC pest officer) and I both suggested that sniffer dogs might be a solution. Following this, we were able to convince QMDC to purchase two Springer Spaniels, "Sophie" and "Rocky", from Steve Austin. Steve trained the dogs and handlers that were involved in removing rabbits from Macquarie Island.

The sniffer dogs are so valuable—not only for finding fox dens but also for community engagement. Tom and I were often on the front page of the local newspapers, not because we were good at fox control but because we had sniffer dogs. Often, people we met while searching for foxes already knew our dogs' names from a newspaper article or TV news.

I recently witnessed some truly successful feral cat controllers. These were Aboriginal ladies in the Tanami Desert in Central Australia. They can tell the difference between male and female cat tracks and even distinguish different males and females. The ladies track cats by following their footprints in the heat of summer days for over 20 km on foot. I was invited along to see if detection dogs had the potential to assist the Aboriginal people's feral cat control. The first day it was too hot for Sophie and she could hardly leave her paws on the sand. The next day we started earlier and we followed a lady called Nolia for over 8 km until we captured a cat. I held Sophie back to conserve her energy but when Nolia yelled "Oy Oy" to indicate the cat was close, Sophie rushed forward to put pressure on it. This may have sped up the work by tiring out the cat more rapidly and we concluded the detection dogs might be useful, allowing hunting to be done in cooler weather and in country where the ground is too hard to see footprints. The skills of these people are truly amazing and they have been working with ecologist Rachel Paltridge for 20 years to protect bilby populations and other wildlife. Rachel has gained the trust of the Aboriginal people. There is mutual respect allowing integration of traditional Aboriginal hunting with science to help manage the desert wildlife.

The final rabbit battlefield in Queensland is to prevent rabbits from establishing damaging populations within the Darling Downs-Moreton Rabbit Board (DDMRB) area in the south of the state. A rabbit-proof fence constructed in 1893 had protected this area from the wave of rabbits that spread from NSW prior to 1950 when Myxomatosis was released. However, during the last 15 years the number of incursions has increased. Will Dobbie, my old friend from Central Australia, worked as the Compliance Coordinator for the DDMRB for over 4 years, organising destruction of key breeding places and trying to encourage landholders to control rabbits. When Will left to return to Alice Springs to work on dingoes, I replaced him to continue his good work.

We conducted a survey of over 1400 landholders, visiting and talking to the residents and inspecting properties for rabbit breeding places. The focus was on understanding distribution and abundance. We located four important breeding places that were previously unknown and would have been a source of reinvasion. In addition to improving knowledge of where the rabbits are we took a large step towards community engagement. Community trust in the Board has undoubtedly been improved. One landholder said, "I didn't think anyone cared". Talking to all these landholders and showing them that they are not being singled out and that something useful is being done seemed to be received very well.

After only 7 months working with the DDMRB I was offered a job I couldn't refuse. Well, I could have but I didn't. As a Research Fellow at the University of Southern Queensland I will continue doing science that is useful as well as having time to publish the important things that should be passed on to others. I hope to continue helping to protect Australia from vertebrate pests for at least a few more years and ensure that others benefit from my experience.

All these different projects that I have been involved in have shown that there is a risk to implementing successful control if the human element is overlooked. Technical-minded people might think that the human stuff is a waste of time. But my experience suggests that if scientists start consulting at the beginning—talking to people and bringing them on board—they can become part of the solution. Scientists need to quantify successes and failures, so we can understand the reasons for success or failure and make better-informed decisions the next time. The way I see it, science helps us get closer to the truth, and using scientific methods to communicate about the issue is an essential part of building trust with community members. Once trust is developed, the special skills of the community members can assist to better manage the pests.

Managing a pest in the absence of good knowledge of distribution and abundance and without measuring impact is like going into a war not knowing how many of the enemy there are, where they are and how strong they are or whether the enemy is in fact an enemy—you won't know whether you've won or lost. That's what I believe often happens with invasive animal management in Australia. People think: "We just want to get out there and do something". But they haven't figured out why they're doing it or where they're going to have the most impact. These questions and answers are vital for successful pest management. Pest management is not just a "people problem" but the people part is certainly as important as the other parts.

References

Dobbie, W. R., Berman D. McK., & Braysher, M. L. (1993). Managing vertebrate pests: Feral horses. *PDF book*. http://www.pestsmart.org.au/managing-vertebrate-pests-feral-horses-2/. Accessed December 11, 2017.

SkyVisuals. (2013, February 5). Brumby: Horse Run Wild. *YouTube video file*. https://www.youtube.com/watch?v=zgv00lndzQ0. Accessed December 11, 2017.

Chapter 7
Practitioner Profile (Brett Carlsson): "The Dogs Are There and the Tools Are There—We Just Need to Work Out the People."

Abstract Can growing up in the Australian bush be an asset to your career? In Brett Carlsson's case, it certainly helps. Brett engages landholders in such a way that puts decision-making back in their hands. He cultivates trust, not through skills he learned in a classroom, but through years of on-ground experience. Bonding with farmers over a shared appreciation for the land allows for open and honest discussion, and helps Brett to understand their situation on a deeper level. As landholders begin to feel a sense of ownership of an issue, they come together and create local community committees. As a result, the community is able to negotiate with the local government with an organised voice. Often taking calls late into the night, Brett is convinced that his quality of relationships with landholders should be a measure of success. His ability to blur the lines between personal and professional relationships has created an important foundation for long-term community engagement and problem-solving in Queensland.

Pest management is not a job for me; it's something I just do. It sounds cliched, but it's true. It's a real passion of mine to learn more about pest animals and how they operate.

As a child, I was aware of wild dog management. I spent a lot of time around older gentlemen of the bush back then, and still do now. I spent significant time in the far southwest of Queensland as a kid on holidays and got experience in dog management there—in pest management in general, I guess. A lot of my weekends and holidays were taken up with managing pests, from a bit of dog trapping to catching feral pigs and spraying weeds. Now, I'm lucky enough to get paid to use the knowledge and skills I've learned from my hobby. I didn't go and study at university; it's practical knowledge and experience that has got me where I am, although I have since done a course and a qualification that says I know what I'm talking about. One thing that the study pressed home to me was the importance of focusing on the impacts of the pest, rather than the pest itself.

My father introduced me to pest management. He had a real awareness of the natural environment because he grew up in the bush. I wouldn't call him a conser-

© Springer Nature Singapore Pte Ltd. 2019
T. M. Howard et al., *Community Pest Management in Practice*,
https://doi.org/10.1007/978-981-13-2742-1_7

vationist, but he instilled an interest in reducing the impacts of pest animals on the environment. I've tried to portray that passion to my children, and now they're just as interested as I am. I've got a 14-year-old boy that can set a pretty handy pig trap and run around a few feral bulls on a motorbike. He's going all right so far—well, he's still in one piece!

I trained as a mechanic after I left secondary school. A few years after I completed my trade, I moved to Longreach in Central West Queensland and started thinking "I need a bit of a change". I loved environmental and natural resource management work, so about 15 years ago, I applied for a job as a Ranger with Queensland National Parks in Longreach. I got to spend a lot of time out on Western Queensland parks and learned a hell of a lot of practical skills in natural resource management.

Then, I went to work for the Queensland Government at Biosecurity Queensland in their pest management division. I wanted to start moving myself up, as a personal development strategy. After that, I was lucky enough to get a position as Pest Management Coordinator with a natural resource management group in Longreach. That's where I really learned my craft in regards to project management and the planning stages of pest management. When the funding for that job ran out, I moved up to Mareeba in Far North Queensland, after a brief stint in Adelaide River in the Northern Territory, working on Gamba Grass control. Then, I worked for the Queensland Government for a while. I still live in Mareeba, but now I work in the central west area as the Queensland Wild Dog Coordinator for AgForce Queensland—they're the industry body representing Queensland's rural producers.

Now, my job is mainly administrative—setting up, coordinating and working with committees; organising programs; and promoting and getting pest management science and research out to people on the ground. I work directly with landholders at the community committee level. Several years back, local governments in Queensland took a lead role in organising and implementing dog control baiting programs and unfortunately the landholders lost some ownership of the problem. I think the lack of ownership contributed to a lack of participation in programs. Like when the local government wanted landholders to step up and provide their own meat for baits, some of the stubborn landholders said, "Well if you want me to do it now and you're not going to do it for me, I'm not going to bother". Others thought, "If you're not going to pay for the plane to drop the baits, I'm not wasting a day of my time driving around throwing them out the car window". Common sense says if your sheep are getting eaten every night, you drive every day putting baits out, but the stubborn brain doesn't work like that.

AgForce decided to submit a funding proposal to Australian Wool Innovation for a Coordinator to be employed and form local committees with affected landholders. The committees put the control back in the hands of the landholders. Local government were saying, "Well, why do we have to make the decisions on when, where and how dog control happens when the landholders are seeing the dogs every day and seeing what the impacts are? They need to be making the decisions". The key driver for the local community committee structure is to get decision-making back in the hands of the landholder.

The key to every good committee is finding the local champion or the local person who really is passionate enough to run it and keep it going. It sounds like a cliche but it's true. Some people were a little bit apprehensive at first. A lot of them told me it was never going to work, that I was wasting my time; but there's nothing like telling me I can't do something to get me fired up! And now, landholders have taken over the leadership of the committees I've established.

These committees have been officially endorsed as a committee of local government, so they can make budget recommendations to their local council and make decisions on behalf of the council. That's a big step, considering this type of relationship didn't exist between local government and landholders in Queensland just a few years ago. We have Wild Dog Management Plans in place that show we're not just a temporary group that's going to come in and tell you what to do and walk away. This has helped build trust between local government and landholders, so instead of using local government as a bashing-board and saying "it's their problem", there's a lot more respect.

There's still a lot of producers that disagree with local government policies, and that's just the way it is, but I think the committees have enhanced the trust and communication between local government and landholders because they've found that local governments do listen; that if landholders turn up with a reasonable voice and not an argument, local government will sit and listen. If both groups turn up yelling—and both local government and landholders like an argument—it just doesn't work. I reckon if you turn up and listen to the opinions of others, then people will listen to you.

The Queensland Wild Dog Coordinator project has also set up a regional committee across a bigger geographical area called the Western Queensland DogWatch Committee. This has been operating as an official committee for over 12 months now. Before that, I was getting the landholders together on a needs basis, but I could see the need to pull them together in a more formal arrangement. I didn't want it to be my idea, so I sort of led them into thinking that running a meeting on a regular basis would be good. I didn't think it was going to float for a while. I thought we might get one or two of these guys coming until we worked out the baiting calendar, and then it would die off. Now, we're well over 12 months in and the first Tuesday of every month, we meet for a teleconference. The meetings go for a minimum of 2 h of a night-time. There are 14 members on it, and we never had any less than seven turn up.

The DogWatch committee has been a real success, a real step forward, because landholders across the region are now communicating with each other, which never used to happen. An example from the last baiting program is that one shire was running out of meat, so they rang the people from their neighbouring shire and said, "Hey, have you guys got any spare meat?" They said, "Yeah, we've got half a tonne left over, do you want some?" So they bought it off them. No way in the world that would have happened three years ago. There's a sort of "mateship" I suppose, between the committee members now. The DogWatch Committee now also has been invited to provide a member to sit on the Ministerial Advisory Group—QDOG.

Sometimes in my role, I am a bit of a counsellor, offering landholders a bit of moral support. A lot of the committee Chairpersons rings me up for someone to talk to and ask, "What are we going to do here?" I throw a few options back at them and say, "Give this a go; what do you reckon about this?" Vice versa, when I have some issues with something, I'll ring them. Sometimes we just talk about life in general and dogs don't get a mention. It's a personal relationship as well as professional one, and that's my management style. I get along with some of the Chairpersons better than others, but that's just human nature. I can also fairly say that all of the Chairs know that they can ring me anytime—and they do, quite regularly—at any old time of the night. It's nothing for my phone to be ringing at 8:30 pm for a chat about dogs. I answer it because that's my job. I know they appreciate that and they tell me that they appreciate it, so it's worth the effort. A real indicator of success for me is how well landholders and committees communicate with me, or how often they ask me to attend meetings or ask for my assistance or knowledge.

Greg Mifsud, the National Wild Dog Program Facilitator (Chap. 13), has been a great colleague and critical sounding board for me. We talk a couple of times a week, sometimes about his issues and sometimes about mine. He also rings me to talk about things that are happening in different areas. It allows me to give my committees and landholders a bit of an insight into what's happening in the rest of Australia. Greg has come to some of my community committee discussions and planning meetings—people really appreciate his perspective of what's happening on the other side of the border, or the other side of the country. The committees are really starting to look outside the square now because a lot of them are a bit desperate in regards to the dog problem, so they want to hear what's happening in Victoria, Western Australia or the Northern Territory. I definitely get value out of that broader knowledge as well.

I've had to educate people about my role. At one meeting, I got run down big time by one of the fiery landholders, who I now get along well with. He stood up and told me that I was a waste of space and my wages were a waste of money that they should be spent on traps and baits and whatever else. People didn't understand the role that the coordinators play. Now, he understands what I do and how I can be an advantage, by pulling together individual committees and the regional committee. I also have input at the ministerial level and sit on committees to represent landholders at that level. The landholders feel really empowered with me representing them at that ministerial level and on these other committees—they're getting a voice.

Not to blow my own trumpet, but if I didn't have the practical knowledge, I've got about pest management—if I just had the science knowledge—I don't think I would be as successful as I am in my role because landholders soon find out whether you know what you're talking about. Plus, I know a lot of the landholders socially from living in Longreach for 10 years before I moved to Mareeba. If I had never lived in Longreach and I was just a Mareeba guy going down to help them put their control programs together and coordinate aerial baiting, I would not be well received at all. But, because I lived in the region for 10 years, I know their plight, I understand their position. To me, it is critical to how my role works; if I hadn't lived in Longreach, this project wouldn't have been successful. I guess that's why they gave me the job.

What makes my job easier is my ability to assimilate and communicate with the landholder on the ground. That's a really big thing. Several landholders have even told me that they value people who can communicate with them and understand what they're saying. As I said, I've spent a bit of time over the years on stations, so I know the sheep and cattle industries. I don't know how many times I've pulled up to go and talk to someone about pest management, and first I've had to help them unload cattle off the truck or finish branding in the yards. Once I show them I can do that practical side of their business, then we sit down and I show them the pest management business. I'm a firm believer that you can put the most knowledgeable person in a position but if they can't understand the people they're working with, it's never going to work. I'm not working with sheep and cattle every day, I'm working with pests; but if I don't understand the landholder's situation or I can't sit down at the table and have a discussion with them about what the cattle prices are doing, or how mustering went last year, it can make it difficult to promote management of pests. If I can't have that honest, knowledgeable conversation with him, then he's not going to embrace me. That's just human nature, to connect with people that understand where you're coming from.

The practical knowledge has also assisted me to provide feedback to researchers on the practical side of pest management. I love the science part of the work—that's been a personal winner for me. I love reading about new technologies and how they might help, and then promoting them to people on the ground. That's a big part of the job for me.

I've just done a monitoring and evaluation program for my whole project. I've also got a monitoring and evaluation program that I want to implement for all my committees as well, to get them to start collecting a lot more data on the impacts of wild dogs. I'll be honest with you—and this is pest management across the board—most landholders *have* got a pest management plan and they *do* keep records, it's just all in their head. Some of these old farmers have got great memories; they don't forget anything. They'll say, "Yeah, well back in 1978 I remember this dog that came through, and that was the first dog I saw that year". Station diaries are great, landholders write a lot more in their station diary than they really let on. I think if you really dug into them and filtered out all the other personal notes, you'd probably get some handy information.

In the past 14 years, there's been a real generational change happening out here in Central West Queensland. Some of the young ones that went away to boarding school and university are getting to 35 or 45 years of age and coming back. They're a lot more tech savvy than their parents, so we can actually send emails and get emails back. They're also a lot more aware of national resource management issues, like sustainable grazing. For example, I went out to a small town right on the border of Queensland and the Northern Territory to promote weed control. One of the younger gentlemen said to me, "Mate, I understand we've got this weed here—give me two months. The old fella is retiring and heading into town to live. As soon as he's gone, we're onto it". Because his dad thought the weed wasn't a problem, no amount of convincing was going to change dad's mind. As soon as the dad left, the son ended up promoting weed control to a lot of the local government areas out his way. That was the generational thing, saying, "We need to do it how dad used to do it while dad is still here". Of course, there are some that just come back to the place and say, "Well I'm here now, dad can get stuffed! We're going to do it this way".

Our biggest problem with pest management in Western Queensland is the landholder that's doing nothing. When I'm talking to landholders at the committee level, they tell me their gaps are the landholders not doing anything in regards to pest control. There's quite a few out there, unfortunately. My personal view is there are a percentage of producers that don't understand what dog control is, or the measures involved. One guy I spoke to have no idea that the local government runs a baiting program each year. You could look at it and say, "Okay, well what bloody rock are you living under?" But it makes me ask, "Well hang on, are we promoting this well? Is it an awareness problem or are they really just living under a rock? Maybe we're not getting the message out there properly?" There are still some people that think it's just too hard—they'll wave their hands in the air and go, "Well, I don't know what to do so I'm not going to do anything". That's the people side of it, so how can we change those people? There's still an education gap there for people like myself and other facilitators and coordinators to fill.

I don't get a chance to talk to non-compliant producers—the bloke that's doing nothing in regards to baiting, trapping or shooting—because I'm preaching to the converted all the time. I was asked a lot when I first started why I didn't go out and talk one-on-one with people who weren't doing anything. Unfortunately, it isn't part of my role. I think if I went and sat and had a chat to non-compliant producers—a real down to earth chat, not a meeting, not a workshop, not as a compliance guy with the stick out—and said, "Well let's have a look at your station dairy, when did your lambing rates drop?" some of them might realise that there is a problem. Sitting down one-on-one could work. In some places, I'd just get marched to the gate, but in others, they might realise there's a problem.

It's not a 5 min job. Some of those catch-ups might end up being a whole day affair. To be brutally honest, I don't have the time. It's frustrating. As soon as I do it for one, the whole lot are going to know, because now I've got them all talking to each other across half the State on these committees! I don't want to set the precedent. I don't want to go if I can't give the person 100% of my effort, otherwise, I'll build up false hope—that's just my work ethic.

Participation in control programs comes back to prioritisation. For example, feral pigs are a big problem for natural resource management groups because they're an environmental pest. They're not so much a problem for sheep or cattle producers because they're not out eating all their grass like the kangaroos is. That's what affects their bottom line and dollars in their pocket; pigs don't really affect how much they get paid for their stock.

I think we've come a long way with the committee structures. Landholders have said to me, "We're starting to get the people worked out, now we've just go to get the dogs worked out", but that will come with more proactive involvement from the people side of things. The dogs are there and the tools are there—we just need to work out the people problem.

Chapter 8
Practitioner Profile (Barry Davies): "You've Got to Personalise It"

Abstract How can engagement practitioners get past previous misconceptions land-holders hold about the organisation they represent? Barry Davies works to break down fallacies by connecting with farmers on a personal level. Barry's first engagement experience was leading a farming initiative with troubled youth from Perth. After a stint in farming himself, he felt motivated to help people help themselves, and began working for the government. Because approaching farmers as the representative of a government agency often created apprehension, he knew he had to prove himself. Barry was able to break through barriers due to his empathy for the emotional and financial toll that farming can take. He has also implemented an approach he calls "positive engagement"—engagement even when there is not a problem. He points out that building strong relationships when things are going well typically makes addressing problems that arise later much easier. From a long and varied career leading engagement initiative, Barry offers an array of techniques for making the work personal, breaking down social barriers and helping others to do the same.

Pest management is something that my family did as part of our normal farming practice. For us, it was no different to spraying for weeds in the crops, which was also managing a pest. My family used to do a lot of rabbit control, mostly shooting and warren destruction. I grew up around 1080, which is a poison commonly used in baiting for a range of pest animals, including foxes, rabbits, wild dogs and feral pigs. The species of native plants (*gastrolobium*) that contain the same compound as 1080 also grew on our property. A small plant is enough to kill 10 or 15 sheep. When we were kids, we would walk the paddocks after harvesting the wheat, find the suckers that grew up through the wheat crop and pull them out before the sheep went into the paddock.

So I grew up in farming—mainly wheat and sheep in the north-eastern wheat belt of Western Australia (WA). Another regular control effort on the farm was fox control, which was done over a three or 4-month period straight after harvesting and to the point of lambing in early April. Between my family and our neighbours, anywhere between a 100 and 200 foxes would be shot per year. Because my family and our immediate neighbours were the only ones doing any fox control work, there

© Springer Nature Singapore Pte Ltd. 2019
T. M. Howard et al., *Community Pest Management in Practice*,
https://doi.org/10.1007/978-981-13-2742-1_8

were always more foxes coming in. This is what's known as the "sink" effect—where the properties that control are surrounded by those who don't, and as a result, the pest problem continues. The only difference was that our lambing percentages were on average about 10% higher than those that weren't doing any fox control, so it still paid to do it.

I did school through to year 12. My last 2 years of high school were at the WA College of Agriculture at Cunderdin. That education was all about farming practices and husbandry, which I transferred back to the farm when I went home at the end of year 12.

I married in my very early 20s and my wife and I ran our own farm until we sold up the bulk of the property in the early 1990s. We sat down and did a 5-year budget projection and due to falling world markets in grain and wool in 1990, we could see that our viability was reducing and we wouldn't be able to finance our debt. So, we made the difficult decision to sell the family farm. Looking back now, it was the best decision without doubt. I took up a short-term contract with the local shire to supervise 17 unemployed and disadvantaged kids aged between 16- and 20 years-old for a 6-month project that was funded by the Landcare Environmental Action Plan. Many of those kids had no or little family support, and had been moved from Perth out to a country town to give them a new start, as a lot of them had either been sexually or physically abused. That job was a massive eye-opener for me, a really good experience. My challenge was to give these kids some security, some rules and boundaries, while working together to do tree planting, building renovation work and also to try and get them into employment by the end of 6 months. That experience was the start of my career working with people professionally.

You don't need a degree to be able to communicate with people, engage with people and get people working together—it's just about being able to communicate with them. It's as simple as that. I was never shy and that definitely helps in my job working with landholders and community members. For me, engagement is about communication, empathy and personality. I would say I've got a degree in common sense.

I joined the WA Department of Agriculture (DAFWA) as a Biosecurity Officer in the early- to mid-1990s and worked my way up to manage the pest animal section of the invasive species in WA, as a Manager for Regional Biosecurity. That evolved into a program focussed specifically on wild dog control, and I became the Executive Officer for the State Wild Dog Advisory Committee. In this role, I was also an inaugural member of the National Wild Dog Advisory Committee, which oversaw the National Wild Dog Facilitator position. During this time, I did a lot of work with community groups, primarily in the pastoral region of WA, which covers all but the southwest corner of the state. In these roles, I wore two hats: one as the Department's official representative; the other as an engagement specialist that worked with community members to give them the tools they needed to manage pests in their community.

I got a big buzz out of that engagement work, helping farmers to gain more control over what they did on their farm or station. I enjoyed working with landholder groups like the Northern Mallee Declared Species Group (NMDSG), a group of farmers keen to help themselves. The Northern Mallee group has been, and still is, very successful.

The Group Leader is the main driver. He runs his farm very well and his drive rubs off onto his community involvement. He addresses wild dog issues for his own personal benefit as well as a community benefit. The Group Leader and I used to be on the phone weekly. He'd ring me up with an issue, and we'd have a bit of discussion about it. We had a good relationship because we told like it was, just matter of fact.

Sometimes, I gave him ideas that he might not have thought of before. For example, when the group wanted to extend the barrier fence, I gave them tips on how to manage government issues, or talk to the minister. I'd work in the background to help them get their message across, to get them over the line. We had a good relationship—we could ring each other at 8 o'clock at night and have a chat. I wasn't fussed about after-hours phone calls because I was dealing with farmers, and I'd have been wasting my time to ring them up at 2 o'clock in the afternoon.

It does get tiring though, and after a long career with the Department of Agriculture Western Australia (DAFWA), I felt like I needed a break. I had either been a farmer or been working with farmers all of my career. I said to my Director, "I'm leaving. I've run out of energy for what the job requires, I don't think it's fair on me or the people I'm working with". I took a 4-year hiatus and went to work with the Parks and Wildlife agency in Tasmania, which gave me the change that I needed. My role there was Regional Operations Manager, responsible for the north-western third of Tasmania, managing places like Cradle Mountain, the Overland Track and other iconic natural wonders. The role included engagement with community organisations, park users and government departments.

Those 4 years in Tassie helped me recharge my batteries. I wanted to get back to working with farmers again. Thankfully, I've got a very understanding wife who's happy to move around, so when a new community engagement position was advertised in Victoria, I put up my hand and was lucky enough to get the job. I was the Community Engagement Officer in the Victorian Department of Environment, Land, Water & Planning Wild Dog Program. The focus of the role was to build a more positive working relationship between government, industry and farmers to tackle the wild dog issue in Victoria.

In the first 18 months in that role, I put my focus into understanding and engaging the farming community. Because I've been a farmer I understand a lot of the circumstances they live with; it can be hard financially and there are a lot of emotional stresses. Personally, I don't want to be a farmer again, but I love farming as an industry and I like the people that do it. They're salt of the earth people. That job gave me the opportunity to work with them, to help them get a better outcome in managing wild dog impacts in Victoria. Farmers should have the opportunity to farm how they want to farm, with no impediments like wild dogs stopping them. If they want to be sheep farmers, they should be able to be sheep farmers. They obviously need to help themselves too, but I see a role for government to assist where it can.

Individual personalities can be really influential in this work. I look at who's telling the story, what position they hold within the community and what sort of influence they have. I gather this sort of information when talking to individuals one-on-one, at community meetings, or through their use of social or traditional media. One thing I watch out for is people who say what they think others want to hear. This is especially common if they're living and working in a small community, because they may tell a few white lies just to be seen as the good guy. This usually makes the government out to be the party that is doing everything wrong. So that's a challenge to get around. Sometimes, they might think they are the community champion, however, other members of the community don't see them in that role at all.

From the farmer's point of view, any animal living on public land, like wild dogs, are owned by the government, I hear that all the time. They feel that the government should be stopping those animals from getting onto their private property. The common view is that they're "government dogs". Some farmers will sit back and say, "Well, I'm not going to do wild dog control if the government won't". I don't think those people will be farming for much longer, simply due to economics. If you are losing sheep, and therefore income, surely it is in your best interest to do something to address that, rather than waiting for the government to do something?

On the other side are those who recognise that the government is doing what it can. They realise the government does not have an open purse for wild dogs because they also need more hospital beds, more roads and more schools. They realise there's a finite amount that's going to go toward managing dogs and therefore if the farmer—if he or she wants to stay farming—will need to do something themselves. The top shelf option—if you can afford it—is to fence yourself in. There is one fence on the highway going up through Ensay in the East Gippsland region of Victoria which a bloody lizard would have trouble getting through. That landholder went from despair at the number of wild dog attacks and considering getting out of sheep, to not having a dog problem at all since he put in the Rolls Royce of fences. Those farmers who can't afford the top shelf option are more likely to get involved at the community level because they need a whole of landscape control program, compared to a farmer who can rely on his fence.

In the Victorian landscape, there are many little valleys that are about 30 or 40 km—about 20–35 miles—long. The cost of fencing those valleys is prohibitive from a government point of view. If the farmers in those valleys joined up and said, "Okay we're going to join our boundary fences to the crown land, and we're going to link up to create a community fence", that would be a great idea. There's one like that in the north-east of Victoria, it's about 60 or 70 km—about 38–45 miles—long. However, like any community action, an obvious problem with community fences is that each individual landholder is responsible for maintaining their fence, so you only need one weak link not to maintain it.

Even so, there are definite benefits in the community taking ownership of the problem and being given the opportunity to run the program in their area. For an individual landholder, being part of a community group allows them to have input into managing landscape scale problems that affect the whole community. Having a say and not relying solely on government to resolve a problem is pretty empowering. But these groups need local people who are willing to put in the volunteer time to coordinate those efforts. They also need people with good administrative and finance skills to keep the group running smoothly.

From all my experiences as a government employee, there are a couple of issues that always come up when working with the community on wild dog management. The first is that getting farmers to agree is like herding cats. As an ex-farmer myself, I can be brave enough to say that farmers can be very self-centred, with the attitude that "if it doesn't affect me, it doesn't matter". For example, there might be 20 different farmers who have to maintain their own section of a community fence and one of them says, "No I'm going to run cattle in there, so bugger the rest of you". Although these farmers live in a small community, they don't always have the small community at heart. They'll come together if there's a tragedy; if a farmer is injured in the middle of a harvest, the community would get together to take his crop off for his family. That would happen in the drop of a hat. But, in day-to-day life, getting farmers to work together is a lot harder.

The second issue is the decreasing population in some farming communities. In some areas, the landholdings have gotten bigger as family farms are amalgamated into bigger agricultural enterprises. So, there are less people in the community in general. If a town doesn't have sheep farmers, it won't have shearers either, which means there might be 20 less kids at the school. As a result, there are fewer teachers, which lead to the loss of one or two players out of the local netball team. The impact on the community goes on and on. Tackling problems from a sheep farming point of view is not necessarily the best way to do it. Because even the cattle farmers want to make sure the sheep farmers stay and their kids keep going to the school, so there are enough kids in the school to keep it open. These are other messages that can get the wider community engaged in the issue.

Over the years, I've done several professional development courses to improve my community engagement and facilitation skills. In 2000 and 2001, I did a Diploma of Frontline Management. I've had media training multiple times. I've learned that if you're not honest and upfront with the community about what the rules are, what the government can and can't work on, you might get away with it for a while, but not forever. It's important not to make promises that can't be kept, because if you don't deliver on it, then you lose all your credibility. That's the way to lose the community's trust, and rightly so. I believe that if you show integrity, that you're there for the right reasons, you will do well.

A lot of government's community engagement is reactive; it's usually based around a problem. If there is a farmer losing sheep, the government will engage with him because he's in the right headspace to get engaged because he's losing money and he's emotionally attached to the sheep he's been losing.

But when things are going well, the government department may be very poor at giving him a call or going out for a visit and having a coffee. There isn't enough positive engagement. That's something I started to do with a few of the farmers that have previously been difficult—the ones who had been very vocal opponents of government efforts in wild dog control. I made an effort to go and see some of them. I went and saw one who had had a pretty heated meeting with some of the government staff around the kitchen table. He'd disappeared into another room and they weren't sure if he was going to come back with a gun or not. He didn't, thankfully. But that had obviously been a pretty tense situation. I went back and visited him after he hadn't had a wild dog attack for 12 months. I sat around the table having a coffee with him, talking about football, farming practices, shearing and at some stage, we got to talking about dogs. But it was just a visit. It was a good opportunity to for me to go and see him as a human being. He's just a normal bloke.

It was also an opportunity for him to see that the government is people too. We do have a heart! If I happen to be driving past his place and he's fencing on the road, I will just stop and have a chat. If he's got some stock in the yard and he's drafting sheep, I will jump in the pens and give him a hand. What better way can you be seen as being a normal person, a human being? To be successful in community engagement, you've got to personalise it. That farmer's going to have a wild dog attack again at some stage—there's no doubt about that. But hopefully that visit, while things are positive, will help him see the government challenges when a dog attack does happen again.

In my role as Community Engagement Officer for the wild dog program, I ran a program of farmer community field days across Victoria. The best place to have these events is on a farm or in the local fire brigade shed, rather than in the government office with a shiny bum like myself standing at the front saying, "This is what we've done, this is how great the government is". I trained up the Wild Dog Control Officers to run these field days—the farmers hold these control officers in very high regard because they're the ones who help them when they're losing stock.

It took a fair bit of prep work to get the wild dog controllers into a position where they were comfortable to stand up in front of people and talk to them. A big part of the field days is demonstrations on how and where to set traps. The wild dog controllers do that on a daily basis but had never demonstrated it to someone else. I organised for them to get together for some practice runs, with six or seven wild dog controllers in a group, and one pretends the rest were farmers and do the demonstration. Then they would swap over, the next one would come in, and he'd do something different. Each person had their own individual method how they would set a trap, and they suddenly realised, "I've been doing this for 20 years and I've just learned something new!" That was really empowering for them.

One of the wild dog controllers had to present in front of 31 farmers. That's a big effort, to stand up in front of 31 farmers and show your skills in setting traps, talking it through at the same time. It's not just setting the trap; it's also explaining all the preparation work, such as where to put the trap, and explaining to the farmers the thinking behind it. For some controllers, there is a fear that by sharing their skills, they won't be needed anymore. But that's not the case. If more people participate in wild dog control, it can take the pressure off the controllers, which means they can be more proactive in managing wild dogs before they become a problem for the farmers. Once the situation is more stable, then there should be less need for farmers to be continuously controlling wild dogs. It's important that farmers are trained by professionals so they know how to set a trap or lay a bait without the risk of educating wild dogs about how to avoid these controls.

Sharing these skills helps build better relationships between all parties involved in wild dog control. It also empowers farmers to help themselves. Farmers are a proud breed and don't like relying on other people too much. Building a strong rapport with farmers through good communication and engagement are the key to getting the job done.

Chapter 9
Practitioner Profile (Peter Fleming): "What's in It for the Stakeholder?"

Abstract Scientists can sometimes find it difficult to relay their findings to the public, let alone change behaviour on the ground. Peter Fleming tries to reach landholders by reframing issues. He always tries to view situations from the perspective of farmers, and to consider what they can gain from their participation in programs. He also notes sometimes the best way to get a message across is to listen to what others have to say first. For example, at an initial meeting in a community, he devotes most of the time to letting landholders express their views and get frustrations off their chest. He finds that this allows for a more productive second meeting. He also employs a nil-tenure approach, which brings people together across public and private land tenures to address invasive species issues. In this approach, building trust across organisations is crucial. Peter doesn't tell people what they "should" do. Instead, he fosters connections among stakeholders and works collectively to solve issues.

When I was at school, I was interested in biological sciences. My mother had a Bachelor of Science, majoring in Botany. She came from a rural background, even though she lived in Sydney. My father's family was also rural people since the 1860s. Both my mother and father were enquiring type of people, and that quality was encouraged in us kids. That, and growing up in the bush, gave me the space and place to investigate how the physical and biological world worked. We used to wander around the bush at home, trying to sneak up on rabbits and kangaroos and watch birds. My mother got a set of binoculars from my uncle; I used to wander around looking at birds, working out what was there, following a platypus up the river, and the water rats. That gave me a fundamental understanding of what drove biological systems. I didn't realise that I was taking that information in, but it was there in the background, and it influences how I think now.

After I finished school, I didn't have a clue what I wanted to do. There was a real increase in environmental awareness around that time, so I decided I wanted to manage natural resources and work on national parks. I never thought about it as a science career, more a management career. So about ten days before university enrolments closed, I decided I would do Natural Resources at the University of New England (UNE), a regional university in Armidale, New South Wales.

© Springer Nature Singapore Pte Ltd. 2019
T. M. Howard et al., *Community Pest Management in Practice*,
https://doi.org/10.1007/978-981-13-2742-1_9

One of the reasons I decided on the UNE was that my mother had been to uni with some of the academics and staff there—like Jillian Oppenheimer in Arts, and Prof. Beadle, John and Beth Williams in Botany and Jill deBavey in Microbiology—and was involved with others like Peter Jarman and John Burton in Natural Resources. Dad had dealings with Wal Whalley about native and naturalised pastures, which he was an early exponent of. So, I already knew these people or knew of them. The family had also been involved in getting the National Parks Association (a community group interested in establishing national parks and biodiversity conservation), set up at Walcha, just outside Armidale. They'd invite these academics to speak to the Association and we'd go on Association bushwalks together; listening to them speak and going to other places gave me a broader feel for the area and the subject.

My official study focus at the University was Zoology. I was really interested in Ecology, but at the time you could only major in Zoology or Botany. While I was doing my Honours part-time, I got a job working with the Geography Department at the University researching eucalypt dieback, where whole eucalypt forests are affected by a range of stressors and start to die. There were two parts to the job—fieldwork and a landholder survey. With the fieldwork component, I did lots of measurements of eucalypt trees, and looked at aerial photographs comparing old forest landscapes with the current ones. The survey involved working with 20 landholders randomly selected from a grid with UNE in the north-west corner. Doing the landholder survey gave me experience in how to set up a survey and understanding about the theory behind the survey methods. It's all learning that I still use. I really enjoyed that job. When I was doing it I thought, "This is useful stuff that I'm doing here".

All these things come together to help explain where I am now. I finished Honours, finished the job, and was then unemployed for a while. Then, I got a job at Armidale High School as a Farm Assistant. After that, I applied for a job as a Technical Officer (Vertebrate Pests) for the New South Wales Department of Primary Industries (DPI), working out of Glen Innes, about an hour from UNE. I went for the interview on a Friday. On Monday, there was a public service job freeze. That meant they couldn't hire anyone for anything; they couldn't even tell me whether I was the successful applicant! About 12 months later, they advertised again for the same position. The description was exactly the same as the one before. I wrote them a letter saying, "I applied before and got an interview, and nobody's ever told me one way or the other whether I was successful or not, could I please have an interview?" I wasn't too pushy about it; I was polite. They gave me an interview and I got the job. My boss said there were two reasons why I got that job—one was that I was confident enough to talk to the interview panel as a group, and the other was that I demonstrated I could develop relationships with rural people, because of the landholder dieback survey I did for the Geography Department. The first job I did for the DPI was to go and interview landholders about their wild dog issues.

I was Technical Officer (Vertebrate Pests) from 1984 until 1986, and then Livestock Research Officer at Glen Innes until 1994. During that time, I used some of the wild dog research to complete my master's degree. The most interesting component of my master's thesis was the final chapter about the relationships between dog control and the number of kangaroos or wallabies in an area. I spent a lot of time in field sites, walking through them, hoping that sometime in the future I could set up an experiment where I could really investigate those relationships more closely. I had several questions: "What are the impacts of dogs on other animals? What do we do to promote livestock production? What effects do those measures have on the environment?" I had a multi-focus approach, if you like.

In 1994, my Research Officer position was centralised to Orange. Since then, I've done lots of different projects and progressed up the Department ranks.

Around 1998, I got an invitation to write a wild dog book. I tried to be a conciliatory person in the book project because I was working with a few "lone wolves"; people with lots of knowledge on the subject and some firm ideas. At the start, I was very much the junior party but I developed my ability to pull people together and get them working in a common direction. We got the book out in 2001—it was a great experience. Those "lone wolves" all turned out to be good friends.

In 1999, while we were still working on the dog book, I commenced a project with a group of other people in southeast New South Wales looking at wild dog management. The nil-tenure approach came out of that project. The nil-tenure approach is about everyone working together to manage dogs across a region, regardless of what sort of land tenure they hold, whether it's private property or public land. Rob Hunt and Noelene Franklin came up with the nil-tenure phrase and had a common goal—"How do we get everyone to work together?"

The relationship between the landholders and public land managers was a big problem, but the nil-tenure approach really gave us a hook to hang our hat on, because people can understand the concept once the problem is reframed. As soon as you say, "Animals don't recognise boundaries," people can visualise it. It's one of those "ah-ha" moments for a lot of people, and you can express it in simple terms: if animals don't recognise boundaries, we've got to manage them across boundaries; it reframes the problem.

What we were really doing was researching management and how to get everyone to work together. We mostly had public land managers and government departments on our side. The project also had a steering committee to help with community engagement. The steering committee was essential to communicating our message—especially when the research involved doing things some people didn't like, like catching dogs and letting them go as part of our research into dog habits.

We had regular meetings with the steering committee and also some public meetings to tell people what we were doing. We would turn up at meetings and tell our story—it was pretty bad actually, the way we did it. A lot of the time, it was just messy. I was abused in a meeting; I was blamed for the telephone company leaving town, the banks leaving town, all sorts of issues. One lady was so distressed that she dumped everything on me as the government representative, and began to cry in the public meeting. I was dumbfounded but I've learned a lot about handling such situations since then.

I learned that the first meeting is basically letting people get issues off their chest—the first meeting gives people the chance to say their piece so that they feel like they're being heard. A lot of community frustration comes from the feeling of going around and around in circles. People have a meeting, they all get up and say what's been wrong for ten years, and there's no action—nothing happens, and you come back 12 months later and do the whole thing again. You get embedded frustration.

For me, the objective of the first meeting is to get an agreed date for a second meeting where actions will be set. If that agreement is all you get out of that first meeting, it's a success. I also learned what to say in those meetings and how to say it. As a researcher, my job was to come and give people information. The people at the meeting would all just sit back and respectfully say, "That's very interesting, that's great, but it probably doesn't apply in this location because you did your work somewhere else". I realised that you have to frame the research around the question "What's in it for the stakeholder?"—whoever they happen to be. If you can provide information to help them achieve their outcomes, you will gain some credibility. The community is not going to invest in you if you're just there for academic interest—they hate that. Be balanced; frame it for them but don't be afraid to present things that you've found. Don't hide stuff from them. Be honest; that's what you've got to be.

In our field, we've got conservation of dogs on the one hand and dogs causing a problem on the other. They're potentially opposing or mutually exclusive outcomes. If you can present the information across both conservation and control objectives, you can achieve understanding and then work on achieving mutually beneficial outcomes. For example, you say, "Alright guys, these people have a responsibility to conserve dogs and you have a responsibility to get rid of the damn things; how do you marry those up?" Once both parties work out what the other group can and can't do, they don't ask for outcomes that are impossible.

As a scientist, I don't tell them what to do. I say, "Here is the information I have, and the limitations of that information are ...". For instance, I can't say that the results of a trial I did in one place—or the data I got from that trial—are the same for another place. I can't be sure that the data extends beyond the area where I did the work. So, I don't tell people what they *should* do. I tell them that whatever they can do is good, and what the outcomes and impact may be.

One of the problems, which we are only just starting to work out, is that expectations of invasive animal control programs are often unrelated to what actually happens on the ground. For instance, the government might issue a control order for a particular pest, directing what people have to do in regards to it, but most of those orders do not have the capacity to achieve what needs to be done; they are often too limited and not based on science. A pest order suggests that every bait you put out will kill a wild dog. That's just not the case. Trying to limit the amount of baits being put out is the wrong approach. Instead, we should be thinking, "Well, how many baits *will* do the job?" We should be asking, "Do our regulations allow people to do enough to kill wild dogs?" If not, landholders are being set up to fail. That's something that we need to sort out.

Another expectation is that landholders will work together to implement control, but there are a number of reasons why this doesn't always happen. First, individuals might see it as contrary to their enterprise to actually do something about it. For example, they might think wild dogs are a good thing because they keep kangaroo numbers down. In that case, it doesn't matter what the law is, the individual's drivers are bigger. They're focused on providing for their family, not complying with legislation. Second, they see dog control as totally irrelevant to them or they don't think there are any dogs around because they haven't seen them; they've got no evidence that they're not complying with legislation. It's hard to prove that they're not.

We've got a real problem in how we sell our message. I think there's an overly simplistic approach taken by legislators and people responsible for implementing the legislation. It's simplistic because they think it's either carrot or stick—that if people aren't complying then we've got to make them comply and that'll solve all the problems. Well, some people are going to respond to the stick, but other people are still going to tell you to get lost because it's not in their interest to do that, for whatever reason.

A lot of our research is about providing explanations for these sorts of behaviours. We try to connect the dots. For example, we ask, "Okay, if people think that, what happens then?" or, "What's the minimum control measure needed to achieve an outcome?" or, "How do we get our message across about how many baits you need to kill a dog?" If you've got a wild dog carcass and you've put out baits that are now missing, it's a lot easier to demonstrate the success of the baiting control measure than if you haven't got a carcass. It's about human perception; seeing is believing.

One thing I haven't worked out is how you convince people who refuse to do invasive animal control to take action—especially when you're an outsider. You can't come into a community and put pressure on a local person, particularly in a public forum. You can't do that because of two reasons: one, it's not a nice thing to do and two, from a purely pragmatic point of view, and you put the community against you if you do that. Even if that person is regarded as a bit of a nitwit, they're the community's nitwit.

Researchers and scientists need to *sell* a message to the community, rather than *tell* them what to do. Not to impose, use the word "should", or tell people what their legal responsibilities are; that automatically alienates those that don't feel included in the community or those who don't want anybody telling them what to do. We've just done some job interviews to find a "people person" rather than a "process person", somebody that can be flexible enough to respond to a group of people. The staff we are looking for have to be polite, they have to be able to listen and synthesise. Not just listen and say nothing—they've got to be able to check that their interpretation of what's being said to them is what people are actually saying. Communication is about messages between the communicator and the recipient, so they have to keep checking that relationship all the time, just checking that the messages coming back to them from the group are those that are being sent.

I try to look for those "people qualities" when I interview people for a job, as I just did, by asking questions, by seeing *how* interviewees respond to questions, not *what* they said so much. I'm not interested in how polished a performance is—I'm interested in how a person distils information and communicates it to the stakeholders. I look at which bits of information they present, and how well they stay on message when they're under pressure in that interview situation. The last thing you need is somebody that goes out there and blows his or her own trumpet. You don't want somebody like that in a public position because they are the type of person that's likely to tell people what they should do and what their problem is.

Reflecting on my experiences has been the way I've learnt about community engagement. I don't have any formal engagement training, and it's definitely been a limitation. I've just gone along and taken it in by osmosis; worked out what doesn't work from reactions and asking, "What did I do wrong there?"

Chapter 10
Practitioner Profile (Matt Gentle): "People Are Key to the Solution."

Abstract Information can be the key to addressing some of the most challenging natural resource management issues, but is all information equal? In his role as a zoologist, Matt often has to interface with community members and local councils. He makes it clear that, in addition to the scientific facts, the beliefs and perceptions of community members have significant bearing on how issues are defined and addressed. Often putting aside his own ideas and scientific interests, he strives to work in the best interest of the community members; in his view, this is the compromise of a public servant. His sincere desire to understand invasive species issues from the perspective of those who experience them day-to-day informs Matt's approach to engagement with landholders and other stakeholders.

I have always been interested in how things work—a natural curiosity and thirst for knowledge. I grew up on a small hobby farm where I was always outdoors, swimming, fishing, camping and so on. Mum was the local librarian and encouraged me to read. Certain teachers at school encouraged me to think about *why* things happen, rather than just describing *what* happens. Science and geography were favourites. Our families were regular campers and bushwalkers when I was a kid, and I developed a love for the outdoors as a result.

I didn't know I was heading towards a career as a pest animal scientist until the third year of my university degree, a Bachelor of Applied Science, Natural Systems and Wildlife Management, when I completed a one semester industry placement with the New South Wales (NSW) Government Department of Agriculture. This really nurtured my interest in pest animal ecology. My project focussed on movement ecology of feral goats. Peter Fleming, a research scientist (Chap. 9) and John Tracey, a technical officer, were my supervisors and they both worked in the department. We were studying feral goats, mustering goats, attaching tracking collars, releasing them and then observing their behaviour and movements. The objective was ultimately to improve management of feral goats in this region. I loved it! It was exciting, and I was working in a team with a great bunch of people.

When I graduated, I knew I wanted to be involved in research, but I wasn't sure whether to embark on postgraduate studies or get a job. I was lucky enough to be

offered a job as a Technical Officer supporting research scientists with the Vertebrate Pest Research Unit at NSW Agriculture, so that took care of that decision. I
held the group I worked with in very high regard. They were open to collaborating
with others, employed plenty of staff and made things happen. They encouraged
an enthusiasm for learning and getting things done. They supported their staff to
undertake postgraduate study, and worked really well as a team. I think that's what
makes or breaks a workplace; it's not the walls, or the location, it's the people. After
a while, I mentioned to research scientists Steve McLeod and Glen Saunders that
I was interested in studying for a Ph.D. They said, "Well, we've got a project on
foxes if you're interested?" I said, "That sounds pretty good", so I ended up doing
postgraduate studies after all. My research focussed on improving fox management
through poison baiting, and involved some very interesting work.

While writing my Ph.D. thesis, I started a 3-year contract as a zoologist with the
Queensland Department of Natural Resources and Mines. I'm still there, and now
I'm a senior zoologist with the Queensland Department of Agriculture and Fisheries.
I lead and undertake research projects on pest animals. My current work focuses on
peri-urban wild dogs, but I'm also working on several other projects, including feral
pigs, cats and foxes. Overall, you could say I am a scientist researching pest animals. I
try to work on applied projects that I think are important and highly relevant. I know
everyone thinks what they do is important, but our research unit aims to deliver
applied outcomes that are directly useful to the end user. We investigate pests, their
impacts and ways to reduce these impacts and how to improve control effectiveness.

My main interaction with the public is through data collection and presenting
results. This may involve collaborations, fieldwork or collating data from other
sources. The process requires engaging with many different people. For example, to
organise a field trip, I have to work with land managers and other authorities who
might need to approve the data collection. It might involve interacting with different
people when going to collect data in the field, or with people who have already collected data that we would like to use, or with colleagues working on other projects.
Determining the priority of research projects also involves much communication and
interactions with many different people.

The engagement processes my team uses are different from those typically used to
get people involved in vertebrate pest control. We're not out there trying to persuade
people to take a particular action. We gather ideas from discussions with people, not
just facts; we collect information that provides insights into how people think, about
what people perceive and believe the truth to be. For us, it's important to understand
people's needs and what we can do to help. Research would be irrelevant otherwise.
After we collect and analyse the data, we communicate our results through meetings,
forums and conversations. We present the work we're doing and ask people for their
thoughts or feedback. Other projects can come from these discussions. The peri-urban
wild dog project is a good example. There, the South-East Queensland Pest Advisory
Forum provided a list of things we should research about wild dogs in peri-urban
areas. The Forum is where people such as Land Protection Officers, Local Governments, Researchers and Community groups meet to discuss the hot topics concerning

pest management across South-East Queensland; their advice helped us identify the gap in knowledge about peri-urban wild dog research that we could help fill.

Communicating about our work is very important. If people don't support it and we can't demonstrate to others that our work is useful, then we're not really doing our job well. We might be doing brilliant work, but if we don't communicate it then it doesn't get a result. For example, I was in South-East Queensland last year presenting at some wild dog workshops. I was discussing 1080 poison, trying to clear up some myths about what it can and can't do. A lot of the myths are effectively rubbish, and not based on facts, but a lot of the people at the workshop believed them. It's hard to shift people from their beliefs. They often say, "Oh well, that's propaganda" or, "That's the government covering things up again".

I don't think I'm particularly good at communicating about the results; I could be a lot better. I try to show too many facts. I try to explain results in different ways, or I might use analogies, or ask what information they need to help them understand the issue or the solution. I try to hit them with the logic of it, but sometimes it just doesn't work. I always thought that truth was power and if I had the truth then I could change the world, but it just doesn't work like that. Ultimately, what people believe is what drives their actions and truths are only one component of what forms beliefs.

Another barrier to communicating the message can come from acting on "second-hand" information. I think gathering more first-hand information from the target community is the way to go. Although a lot of the project ideas we work on come directly from people or local councils, often we are only receiving second-hand information about the issue. It's so important to know what people think is the real underlying issue. I don't want to do research the wrong way, or target the wrong questions. It's much easier to sell a message when it reflects what people are really thinking. For example, if all the councils say, "We have a problem with feral pigs", then I need to ask, "Okay, what is the real problem?" and go from there. I really have to talk to people about what they think the issues are, then they can be part of the solution—then they are the solution!

I reflect on meetings where I've struggled to get the message across, and ask, "How can I do better next time?" I don't think I can ever get the message across 100%, but if I encourage most people to listen, or at least be willing to hear my perspective, then that's a good start. I know there are different ways to communicate that I haven't come across yet, and I'd like to learn a bit more about them. Even though the use of different communication methods might not get everybody on side, it could probably get a few more.

Personally, I think people should aim to be persuasive when they speak, but persuasive in terms of the right message. You can be confident and stand up there and talk about something that you think is happening, but that is your perception of reality, and may not match or be suitable for everyone else in the room. It is difficult to reach everyone in the room.

You need to consider the message in terms of what is best for the public good—that's the sort of compromise you need to make when you're a public servant. Within government, you have to be working for the public good, not running off on a personal tangent. You need to be able to sell your work as being important to the public. I aim to do research that people are going to use, so I accept that I have to compromise what I personally see as most important. It's all about the big picture in society, isn't it? What's personally frustrating is when obvious things in the pest management space that could be easily changed, aren't. If it is obvious and people are not doing it, then we need to find out where the barriers are. This is part of the challenge of working in this field.

There's no point in coming up with an idea that is socially unacceptable. In order to get outcomes that people are going to use, pest managers have to consider other things that are going to influence implementation, such as social barriers. We could be stubborn and try and push through, but in the long term, it's not going to work because people aren't going to use our results if we do that. I don't think we can ever completely fill the gaps in pest animal policy with research—there are always too many questions. Sometimes, there will be an "easy" answer to pest problem. When we can't see the problem (or answer) easily, sometimes, we just have to make the best of current knowledge until we have the resources to investigate through further research.

Most of the problems we deal with are people problems—I'm realising this more and more. I also realise that people are the key to the solution. There are a lot of different personalities in the pest management space. For example, the people who attend pest animal conferences and weed conferences have different personalities from landholders in isolated areas, or those in peri-urban areas where you're dealing with a whole spectrum of people—people who love dogs, people who hate dogs and people who trust or distrust institutions. When I was 18, I didn't understand how everyone needs to work together as a team. Now, I realise how important it is, especially given scientists are increasingly working with end users. For example, on one project we had six study sites with ten people at each site. We had to speak to them all before we went out and did our work, and if we didn't get along or if they didn't like us, they could just deny us access and halt the whole program. It's been a bit of a wakeup call. I enjoy working with scientific data because often it's black and white. The people area is a bit "greyer".

Scientists often have the tools and the strategies to implement the scientific measures, but getting people involved is where the real battle is. Sometimes, I just don't have a solution; I just keep working with them, trying to get to a compromise. I've had issues with collaborators who don't really support the work, or think that the way I approach things is wrong. I just keep talking to them, working through it, and we eventually come to an agreement. It is hard to get everyone on the same page, but it helps to understand different communication styles and to identify people's strengths and how everyone can work together.

As a scientist, there will always be things I don't know but I try to keep developing skills in those areas. Communicating the message is just too important. There are great people to work with who are also great at community engagement. For example, our team had done a lot of engagement work with Darren Marshall, Senior Project Officer for Pests and Biodiversity with the Queensland Murray Darling Committee (Chap. 12). Darren's great—a real salesman no doubt, but he has a great product to sell—he engages the community through a collaborative approach to the problem.

We have worked with others like Darren who is great at getting people involved. That has also improved my understanding about the sort of things we need to do, such as selling the message in different ways, not just the old traditional ways, as well as working more with engagement specialists to learn how to get the message across and how to listen to others. Listening to others is important because they have valuable knowledge to share as well.

Over the next 10 years, I think we can dramatically improve the uptake of research findings through increasing our understanding of the barriers, and ways to overcome the barriers to participation in vertebrate pest management. I am keen to keep collaborating with others to engage people as an important component of future research projects.

Chapter 11
Practitioner Profile (Jess Marsh): "It Takes Years to Build Trust and a Day for It to Go"

Abstract Transitioning to a new job can be difficult, especially when you feel that you're starting from scratch. Although she faced challenges when first taking on an engagement role, Jess Marsh's perseverance and willingness to learn from others contributed to her success. Jess depended on the relationships and trust she built with landowners, government officials and co-workers in order to see projects through. Early experiences also showed her the importance of understanding local context. Now, before she starts any new project, Jess looks into what has already been implemented, what worked, what did not work and why it did not work. She recognises that, based on historical context and previous patterns of interaction, the same issue may need to be approached differently in different places. For this reason, she seeks to hear all sides of a story. With a genuine respect for people that transcends her desire for any professional gain, Jess is able to build relationships with landowners that put them at the centre of decision-making in communities.

I grew up on a small farm with my family and that is where I learnt about animals, farming, the value of water, landscapes and neighbourly politics! Studying and working in the environmental and agricultural field was always a given, it just felt right and I wasn't ever interested in anything else. At the university, I did Zoology and Marine Biology. My degree had a strong practical focus and by the end of it, I felt ready to launch into an Honours project. I always thought that science was just research. I didn't think about the people side of it—I didn't think I'd ever be any good at that or need to even think about that. I just wanted to go out in the field, do some research and find out things. Thinking back, my entire university degree barely touched on invasive animal populations in Australia and it definitely didn't touch on the human dimensions of anything!

Family circumstances changed everything and instead of staying at uni in Townsville, I returned home. In 2005, I started working in pest management at the Department of Primary Industries (DPI) in Orange, regional New South Wales (NSW). My first role was very technical—a lot of fieldwork and assisting researchers. The job helped me develop my skills and knowledge in the technical aspects of pest animal management, but it didn't really touch on the human dimensions at all. At

© Springer Nature Singapore Pte Ltd. 2019
T. M. Howard et al., *Community Pest Management in Practice*,
https://doi.org/10.1007/978-981-13-2742-1_11

least that's what I thought at the time. When I look back now, I can see that I was doing "engagement". I was responsible for finding farms suitable for our pest management research, contacting the land managers to make sure they were okay with the work being done, making sure they understood the project and communicating project results to them.

As the funding for the technical job ended, I transitioned to the job I have now, which is mostly a facilitation and human engagement role. I had a discussion with Glen Saunders who was the Lead Research Scientist in the Vertebrate Pest Unit at the time and he said, "You'll be fine, you'll be great". I thought he was crazy. I remember saying to him, "I don't know anything about people!" He obviously knew better than I did or was prepared to take a risk. He had seen me work with different researchers and as part of his team. Glen would have been aware that I was working across different species and with different researchers, and maybe that's why he thought I might be good at the job.

Now, I'm the National Natural Resource Management (NRM) Facilitator for the Invasive Animals Cooperative Research Centre (IACRC). The position was created about 7 or 8 years ago, to fill a gap between invasive animal's research and on-the-ground work in NRM. When I first started, my understanding was that I was to work with the 56 NRM groups operating across Australia. These groups were originally set up in a partnership between the federal and state governments to coordinate NRM activities in their region. The National Facilitator worked to promote best practice management techniques and the research of the IACRC to these groups. Over time, it's become clear to me that the position is much broader than that. I don't just work with the NRM groups—I also work with a range of government departments and authorities, community groups and land managers.

The main aim of the position, as I see it these days, is to try to get people to work together in managing pests. It's quite a broad mandate, and there have never been any major restrictions put on it. In a lot of ways, that's good, but it is also challenging at the same time, because it makes it such a big job and it can feel quite reactive at times.

To be honest, I thought it would be a short-term thing, but I've been in the role for nearly 6 years now. In the beginning, it was terrifying—partly because I thought I'd never worked with people before, but also because the handover from the old Facilitator was terrible. All he really said in the handover was, "You'll be fine. You'll hate this job less than I did". It would have been good to have a handover from someone who liked the job. They could have shared their knowledge and effective approaches with me. I basically started the job from scratch.

The position used to be called NRM Liaison Officer. I remember the old Facilitator saying in the handover that he was a Liaison Officer, *not* an Engagement Officer. This was my first insight into what it actually meant to "engage"—what was the difference between liaison and engagement? I think he thought he was doing his job by purely contacting and informing people—doing that very basic level of engagement.

Mike Braysher, a lecturer and Adjunct Professor at University of Canberra, probably had the most to do with establishing the position and helping me develop the position over time. In the early days of the job, he was the person I'd ring and say,

"Oh my gosh, what am I doing?" He was probably the only person that knew what the position was supposed to do and who could guide me through it. If it wasn't for him, I'm not sure which direction I would have gone in. I would have got there eventually, I guess. It helped that I did a Diploma in Community Engagement that he used to run out of Canberra University; that reinforced what I was doing and that I did know the principles of engagement.

At the start of the job, I spent 6–12 months trying to establish myself in the job. I created the "NRM Notes" newsletter and did a lot of mail outs and direct contact introducing myself, sharing links to information, letting people know that I could help them with their planning or engagement. It was hard to set boundaries with people at the start. Sometimes I'd say, "Well, I can help you with that" and they'd say, "Oh, but what about this?" I learned the hard way about taking too much on. It's still one of the hardest things about the job—do you pick a few regions and do something specific or do you broadcast information to lots of people without doing anything specific? I try and cover both those bases without running myself into the ground.

About 6 months into the job, I put in a tender for a team to develop a pest management plan for one of the NRM groups. I was terrified that we'd get it; I still thought I didn't know what I was doing. Again, Mike was there the whole time. He had developed the pest plan process, a specific approach to pest planning with the community, and had also recently run workshops in the western division of NSW. My team got the contract and, even though it was scary, I learned so much about the process of engagement as we developed that plan. We hired a professional workshop facilitator and ran a series of community workshops across the group's catchment region. We ran the workshops over a year-and-a-half, and then developed the information from these meetings into recommendations and priorities for that region.

The process laid the foundations for a good relationship with that NRM group and all the government agencies who were represented at each workshop. I learned what it took to be able to get to the stage of calling people and saying hello, catching up and then working together on projects. I realised that those valuable relationships take a lot of time to build, because you can't build relationships without trust and trust takes time to develop.

Even though the process was terrifying, it was also valuable and rewarding—even now, because people still refer to that plan. They've used some of the recommendations in the plan to make successful funding applications. That's nice to see, because a lot of plans don't get used. I learned so much from the workshop facilitator we used for the workshops. I still think about it, 6 years later. A lot of the time, he showed me things indirectly—just watching him with people weres valuable. I also learned a lot from spending time with people, seeing how they worked and responded to different cues and body language. I look back on that first planning project and think if I could do it all again, I wouldn't do it any differently. It gave me a bit of confidence and reinforced the message that people are key to making progress.

It's hard to gauge whether or not I've been successful in the role, because it is so broad and engagement itself is difficult to measure. It's hard to document "engagement success". It's so much easier to measure specific indicators, such as the number of pest animals killed or landholders at field days. I disagree with just using those measures. Personally, if I know that I've helped a group or individual with a problem, then that's success. For me, it's about the whole picture, including the depth of engagement. I can send out a million flyers tomorrow and say "Job done", or a million people can go to a field day, but have those things actually changed people's thinking or behaviour? Unfortunately, most reporting requirements are still based on those measurements alone, maybe because it is much harder to gauge if people are engaged. Anything to do with what people think or how they feel is hard to measure—everyone is just so different.

I have to be very realistic in this job. Sometimes, nothing I do is going to change people's thinking or behaviour. Accepting that is part of being realistic and understanding where I can be most effective. I think there is a lot of pressure on engagement practitioners. I used to get quite anxious and stressed when people would be proactive for six months of the year, reduce the pest species impact and then stop. I used to think that that was my fault—that I was failing at my job. These days I know there is only so much I can do. You can give someone support and every tool under the sun to manage pests on their land but whether they do it or not is still their decision. Understanding these decisions, the background behind them and how to approach them is really important. I have to be realistic, accept different views and then figure out how to work with them.

It's helpful to have a good understanding of community engagement principles, but these principles need to be applied with a certain amount of flexibility and caution. Every problem and solution has a different history. This history can be positive or negative. Before I go out into an area, I try to look into what has already been tried, tested or implemented, why it did or didn't work, and the individuals that may have created conflict or driven the project. For example, I might get two calls about rabbits and both people might have the same issue, but the approach I take with each person is going to be different because they are different people. That helps me understand who I'm working with and what we're working towards.

Personally, I think face-to-face conversations are the most valuable of all communication methods. I get a lot out of meeting people and it takes less time to become familiar with people you can see and respond to. I pick up on a lot of their characteristics, approaches and body language. I've also learned over time that I cannot and should not take everyone's word straight away. I've listened to quite outspoken people and made the mistake of assuming that's how it's happened, only to find out that that's not how it's happened at all. There are always at least two sides to the story. I've learned that someone in my position needs to be open-minded, prepared to listen, and seek more than one opinion. I've got to be careful not to take sides. Most of the time, people are happy to give me information. I just need to make sure that it's the correct information, especially if it's about other people in that area, because if I'm seen to be on one side then I'm instantly offside with others. That would make my job almost impossible. Trust develops over time, and I know when I'm going into a new area that I have to be very cautious.

Pest management is not just about on-ground pest control techniques—it needs people to put it into action. The techniques and control options are not the only answer; it's also about how those tools are best used. Sometimes people are difficult. Sometimes, inaction might be because of a past personal relationship. Sometimes, it can just come down to a lack of awareness because there are a lot of myths out there. That's okay—it just means I have to change my approach. Most of the time it comes down to a misunderstanding. My approach is to go back to each group or person to really identify the problem. Having those conversations may seem like taking a step back, but for some people that's what I've got to do. I ask people what they need and how they see it going forward. It might be as simple as hearing from a researcher that this is the best approach, or from an external person that's not from their area. And sometimes, the best-managed pest control approaches can fall over because key people leave or become overwhelmed. Working with people and trying to move through these challenges is vitally important for facilitators like me. But sometimes, I've just got to walk away and accept it's not going to happen.

Even though I have been working in this field for a while now, I still make the effort to talk to people who have a great depth of knowledge across many fields and areas of research. I also communicate with other facilitators regularly because even though we have similar roles, we approach things very differently. What I get from other facilitators is the understanding of what the position requires, new approaches and emotional support—which is really important! I feel like I have friends in this area as well. I think that being friendly is what I have to do for this job, but it also just comes naturally to me. I still talk to the facilitator of the first planning workshops, to debrief on heated issues and get advice from him about workshop facilitation. It's good to have emotional, factual and practical facilitation support.

I honestly think that, regardless of formal training, effective engagement depends on personality. Formal training helps, but the bottom line is the engagement practitioner's personality—you have to click with other people and find ways to relate. On top of personality, you have to be very patient and you must always listen. There is little acknowledgement of how much time it takes to do engagement well and how much depends on personal relationships and trust. The way I see it, if I have built trust with someone, I can call them up and we pick up where we left off. If I say, "let's take this action", they're not as cautious.

Trust builds and becomes stronger and over time. Trust works not just between two people, but it spreads to others as well. For example, if I trust you and you trust your neighbour, then I will also trust your neighbour. Because of those connections, you have to be very protective of your relationships. I've worked hard for a long time to build up relationships and a reputation, and it can go bad quickly.

The longer I'm in this job, the more I invest in relationships. I have learned the hard way that if I'm asking people to give me their time, they need to be updated, included and thanked. I've learned that if I make a mistake then I need to acknowledge that and I need to say sorry. I've learned that building my trust in others and gaining others' trust are both challenging tasks. It takes years to build trust and a day for it to go.

Chapter 12
Practitioner Profile (Darren Marshall): "A Learning Journey for Me and the Community"

Abstract Although the benefits of working together with community members seem evident, there are times when engaged practitioners can become incredibly frustrated with their work; when this happens, they may find refuge and hope in each other. Darren Marshall almost gave up on engagement until a colleague from across the globe was able to reignite his passion. A course in the United States helped Darren bring a new lens to his work with communities. Instead of presenting research, technology and processes as he had in the past, he began to develop research with communities right from the start of projects. Darren no longer separates the biophysical and human dimensions of an issue—to him, they are equally important. He tries to consider all sides of an issue, and has learnt how to gently challenge an idea and how to address the concerns of forceful individuals. Without the whole community's input, an effort cannot be sustained in the long-term. Drawing inspiration from others, Darren himself inspires, teaches, mentors and empowers community members to work together in the most effective ways possible.

I am currently the Senior Regional Officer for Biodiversity and Pest Management working for the Queensland Murray–Darling Committee (QMDC), a Natural Resource Management (NRM) group. Our region covers the entire Murray–Darling Basin catchment of Queensland. I focus particularly on feral predators like wild dogs and feral pigs. Over time, I've come to realise that small-scale control is ineffective when you look at the home range, density and impacts of these animals. The approach now used by QMDC is to work with groups of landholders to coordinate their baiting, trapping and aerial shooting efforts to achieve long-term, landscape-scale outcomes.

The natural environment has always drawn me, even as a young person. I grew up on a small farm in Glen Innes, on the Northern Tablelands of New South Wales (NSW). My mum has worked in the same chemist shop since she was 19 and is still there today. My dad runs an earth moving business, predominantly working for landholders. And they both took care of our family farm, which I still manage with my mum today. As a kid, I really enjoyed spending time with my grandparents and cousins who were from farms close to home. Looking around that landscape, I used to think, "How could you not want to be a farmer?" I was always surrounded by

© Springer Nature Singapore Pte Ltd. 2019
T. M. Howard et al., *Community Pest Management in Practice*,
https://doi.org/10.1007/978-981-13-2742-1_12

cockies—that's what we call farmers—and so I understand the way that they think and operate, perhaps more so than other people who didn't have that experience growing up. Because of that I feel that I can create a really good rapport with them and engage effectively with them.

After I finished high school, I joined the army—it's just what young fellas do, isn't it? It was something constructive to do while I made my life decisions, and it gave me the opportunity to pay for my study. I was in the army for 4 years in total. My first year was full-time—I went through basic training and then I was posted in Brisbane. Being there gave me a taste of city life and confirmed that I didn't want to be a city slicker—I wanted to be outside, on the land. After leaving the army, I decided to do a Bachelor of Applied Science in Environmental Resource Management at Southern Cross University in Lismore, Northern NSW.

There were a lot of greenies, or environmentalists, studying there. I'd come from a farm, pretty naive, and then went through the army, which is pretty strict, and then into this—what a culture shock! It was good for me. I saw the need for a balance between production and biodiversity, and worked out that I really wanted to protect Australia's unique native fauna. I feel like some practitioners are solely focused on agriculture, while other practitioners are solely focused on biodiversity. Those different emphases can lead to very different outcomes when working with the community.

When I left university, I did volunteer work at Springbrook National Park, and Natural Arch National Park in Queensland. When the opportunity of a job came up, constructing walking tracks and rock walls in those national parks and then in Washpool National Park, New South Wales, I jumped at it. That was for about a year.

Next, I became a Technical Officer for a soil mapping project based in St. George, Queensland, with the Department of Natural Resources. When I first drove into St. George, I stopped on the edge of town and thought, "What the hell am I doing, I do not know a soul in this place". Then I met Luke Zohl, who became a great mate and introduced me to the pastime of pig chasing. Here is where all of my worlds seemed to come together. I could be outside, protecting Australia's wildlife, using my military skills shooting and really enjoying my work. Between that and rugby, it did not take long to meet half the town.

My role was to help map the land systems and soils in South West Queensland. I learnt a lot about soils and landscape processes but it didn't really involve people, other than my colleagues. Our only engagement with landholders was to get permission to go onto their properties to take soil cores. We'd describe soil cores and then map the landscape. Landholders would sometimes ask for information or they'd want us to soil test a certain area they may have been concerned with, but because we were in a project that was confined to mapping and the person that I worked for was extremely focused on that outcome, we didn't engage the way I thought we should. That job was challenging from a human dimensions point of view, simply because it always made me feel uneasy when a landholder was good enough to allow us on their property, but we did not work with them on their concerns. I always felt anxious about it.

While I was in St. George, I also met an old fella called John Gray who was a Soil Conservation Extension Officer based in Warwick, South East Queensland. I first came across him at a salinity field day with departmental representatives and landholders in the Goondoola Basin near St. George. He asked me what my ambition was, and I told him that I wanted to be a Land Protection Officer. I remember his words: "Don't be stupid, boy. It's a sunset industry". Then, he grabbed me by the scruff of the neck—rough as guts—and said, "You communicate well with cockies, we want you to come and play a Landcare role with the community in Inglewood. That is your calling". Talking to people has never been a problem for me. I don't flick a switch to turn that on, it's something that I love to do.

I did an interview in front of five landholders and John. I can remember one of the farmers saying, "If that boy has been to university and studied the environment, and his old man's an earth moving contractor, he should be a good balance". So, I got the position of Landcare Coordinator in Inglewood, in southern Queensland, working with landholders to improve environmental and agricultural outcomes in the district. Landcare assists landholders to integrate environmental outcomes into their agricultural enterprises, through field days and bringing experts to share information, helping landholders get grants to improve the environment and organising educational activities. The Landcare role was my first exposure to working with communities; this was where I really learned that you can't address environmental issues on an individual property basis, because environmental issues don't stop at property or even state boundaries.

John was my first mentor and still plays a huge role in my life to this day. One of the main projects we worked on together was with a group of landholders in the Inglewood and Stanthorpe Shires to improve the quality and markets for their wool producing members. It was entirely community driven, and a well-recognised force in the wool industry. We developed a partnership to develop and map a vegetation plan for their entire coverage. They covered 6 sub-catchments, which was well over 30 properties. This is where I learnt a lot about working with the community, because we had a meeting Tuesday, Wednesday and Thursday night every week for 3 months. On the way home from each meeting, there was at least an hour's drive and we would talk about what I did well and what I could improve from that night's meeting. Because it was over such a long period of time, I feel like I learnt a hell of a lot about how to engage with groups of landholders. I learned that you always have to have at least two allies in the room. And I learned other basic things—to consider every landholder's point of view, and never disrespect their opinions; the importance of the participants having ownership of what is trying to be achieved; how to gently challenge people's thoughts and how to bring new ideas to the table. I was told there is no point being an extension officer that doesn't take the group on a learning journey, while also learning along the way.

At that time, the NRM bodies were being created in Queensland for the purpose of channelling funding from the federal and state government to the community to improve environmental outcomes. There was a lot of opportunity around. So in 2003, after 2 years with Landcare, I took a position as Planning Officer with QMDC. My first role was to continue working with John and the wool growers on their environmental management. I worked closely with Tony Gleeson and the Australian Land Management Group (ALMG) to implement Certified Land Management (CLM) with landholders across Australia.

Through mentorship, guidance and support from my two key supervisors at QMDC, Bob Walker and Geoff Penton, over time I took on other roles and responsibilities. I became Regional Coordinator for Feral Animals and Weeds in 2005. I was in charge of a staff of four to eight people, managing budgets, writing community grants, meeting reporting requirements for the government and developing partnerships with key organisational stakeholders. I still do this for feral animals, but not for weeds anymore; the staff management, in particular, takes up so much time. As a result of that partnership building and developing successful projects, commercial opportunities started to present themselves. This was timely, because as state and federal government funding declines, bringing in commercial work has become more and more important.

My responsibilities now include the management and implementation of feral animal commercial contracts. Industries like coal seam gas mining companies contract us to manage feral animals on the properties they own. Their focus is not only on feral animal control but also research to assist the community to improve feral animal management outcomes. The target species are wild dogs, feral pigs, feral cats, and foxes. Over time, other stakeholders have seen value in the work we do. Now, I also have contracts with state government agencies, local governments, industry groups and community groups.

I am always looking to network with and learn from researchers and extension officers in other geographical areas who might be able to offer new insights. In 2012, I travelled with colleagues to the International Wild Pig Symposium in San Antonio, Texas, USA. On that trip, we arranged meetings with researchers at the Kerrville Wildlife Research Centre, Texas; Mississippi State University; and Penn State University. At Penn State, I connected with Ted Alter, Professor of Agricultural, Environmental and Regional Economics. We got on like a house on fire. Ted was already involved with the Invasive Animals Cooperative Research Centre (IARC) in Australia, and we began collaborating. We have never looked back, and as part of this collaboration, I travelled again to Penn State in 2013 to take an intensive short course called Leadership and Community Engagement, for Australian researchers and practitioners in the NRM field.

The Leadership and Community Engagement course really marked a turning point for me; it was the most powerful course I've ever done, hands down. Before the course, I had been about ready to throw the towel in on working with the community. People seemed to block or be in the way of the greater good for everything I wanted to achieve. I was going to purely focus on ecological research—work with the animals and leave the people behind. The connection with Ted reignited my natural ability to engage and work with people just in time. The course was structured in such a way to show how community engagement is actually practiced. Previously, I was trying to bring information, relevant research and effective planning tools *to* the community to assist in addressing the feral animal issues. The concept of developing research *with* community involvement from the beginning is not often practised by research organisations in my experience in Australia, and this changed how I thought I should work. This is where I started to realise that co-created projects that lead to a joint journey, (my own and the communities) with a mix of local experiences and knowledge, as well as research, could be extremely beneficial.

I am always looking for the next step, and the opportunity to complete a professional doctorate (Prof. Doc.) degree through the University of New England (UNE) was something Ted and I talked about during the course. A Prof. Doc. is essentially a Ph.D. that you do at work, centred around the development and implementation of an innovation that will have a real impact or change in the community. It seemed a natural progression for me. I don't see the biophysical research to be any more important than the human dimensions research, or the human dimensions to be any more important than the biophysical. They're both equal. So why do we have so many people who just work on the biophysical aspects of an issue, and so many people that just work on human dimensions aspects of an issue? I want to be the bridge, and I think that my Prof. Doc. research is placing me on track to do that.

My idea came from some research I had helped with, looking at how feral pigs interact with domestic free range piggeries. Using GPS collaring to track the movement of feral pigs, I realised that what people thought and what the feral pigs actually did were often pretty different. The collaring data I gathered with groups was really powerful. A lot of landholders in the study area believed that the feral pigs lived in the national park   which was some distance away—and the pigs would travel every night to eat their crops, and then go back to the national parks to camp. The thought was that if the national park was not there, then there wouldn't be a feral pig problem. It was clear from the collaring data that the pigs did not leave the landholders' properties and they did not once even visit the national park. That data gave the whole community ownership of the issue. It was locally relevant applied research, and they sat up and took notice. As a result, they now coordinate their control efforts. The collaring was the vehicle that changed the community's perception. This example inspires me that what I am doing with my Prof. Doc. is on the right track.

In my Prof. Doc., I am testing an innovation using applied research as the vehicle for engaging the community in different agricultural areas in Australia. Specifically, I am testing to see if landholder participation in a project using GPS tracking collars can build individual and group ownership and commitment to address the feral pig problem. I believe that when ownership and commitment of the community is strong, more effective strategies and management plans to reduce the number of feral pigs can be developed and their agricultural and environmental impacts reduced. In my experience, when community residents work together to address an issue important to them, they are then poised to handle other issues that may confront them going forward.

Looking back at my career so far, I've been lucky that I have always had strong mentors. These include John, Ted, Bob, Geoff and, more recently, Matt Gentle (Chap. 10) and Dave Berman (Chap. 6). A lot of people in my generation think they know a hell of a lot more than they do, and they don't draw upon all the older experienced people out there who are so willing and giving of their time and wisdom. But I've always been lucky enough to be surrounded by experienced people and tried to learn from them; my mentors have all seen community engagement as essential to making any long-term change.

Doing the Leadership and Community Engagement course and working towards the Prof. Doc. has helped me to build my own understanding of engagement. At QMDC, we developed a structure for engagement that has four pillars: regulation, incentives, research and education. Different organisations engage with land managers within one or more of these pillars, but also come with different agendas, beliefs and ways of thinking about their issues. There are many sets of priorities and sometimes they are very different, even conflicting. To create and implement on-ground motivation and change, I need to engage with the land managers, but also stakeholders from the other pillars. Ideally, we're all heading on a journey to achieve a collective goal and not just an individual property goal or, in the case of stakeholders, a broader industry goal.

In theory, legislation can assist collective action by enforcing participation for the few recalcitrant landholders not willing to be involved. However, in practice, this doesn't happen. In Queensland, there is a legal obligation to control class two pests—any feral animal with an established population—under the *Land Protection (Pest and Stock Route Management) Act* 2002 (QLD). In my experience, I have never been involved in or heard of a local government using their authority to enforce feral animal control, in part because they can't prove who "owns" the feral animals. The legislation is often seen as an idle threat, and is described by many land managers as a "toothless tiger". Because legislation isn't enforced, it actually makes it harder to get people to take collective action.

Incentive programs can also decrease landholder's ownership of the problem. If someone comes and aerially shoots feral pigs on a property for free, what stake does that give the landholder? If in 2 years' time feral pigs are a problem again, people will begin to say, "Hey *your* pigs are back. When are you going to get the helicopter back and shoot *your* pigs?" This highlights the importance of community engagement to build ownership of the problem. I believe that if work is continually completed for land managers through grant schemes, or by others, then that is the only time those landholders will act. The question for me is how we build ownership and motivation to inspire communities to take greater responsibility for their own feral animal problems, and to act together to address them.

There are still so many biophysical and human dimension questions that need to be answered. These are not separate in my mind and they need to be crashed together. For example, we need to know the home range of a feral animal species for ecological purposes, but also in order to structure group size for the management of the species. We say that we aim to get 10 or 12 landholders working together, but where does that number come from? Should it be 5 or 6? Should it be 30 or 40? We need to know at what scale our control methods need to be implemented to ensure there is a noticeable impact on the population.

To answer these types of questions, part of my work at QMDC is to create partnerships with different research groups. I work with them to extend their research from the lab to on-ground field trials; they often have small-scaled pen trials of toxins for feral pigs, and when they get their results to a level that can be field tested, that's when we test it with the community at a landscape scale.

Effective community engagement is about having landholders involved from the start, getting their input into the research project and then beginning a joint journey of learning. I've learnt a lot from working with communities, and am still learning how to improve. Originally, I had in my mind that if I could develop really good scientific tools and extension strategies, I could convince landholders with solid research data. But that hasn't worked all the time. I believe the landholders have to be a part of the science; you need them involved from the start of your idea. Then you can build joint ownership and hopefully get some on-ground change, because they're willing to adopt it—because they have a stake in it.

Production impact is one of the hooks that I use to engage the community, and to help them think beyond what's politically hot at the minute. The second hook is biodiversity impact, the third is disease spread and the fourth hook is about being a good supportive neighbour and working together as a community. I have good practical applied biophysical research that complements each of those hooks. I use these different hooks to get landholders thinking beyond their own enterprise or their individual property, to thinking that we're all fighting the same pests. We have mixed farming systems. The ideal dates to target control for different enterprises don't line up, so people are not motivated to work together as one united front. So I ask, "When, ecologically, is that pest at its weakest?" Then I can stop the sheep fella saying, "Okay we only want to control before lambing". Or the chickpea fella saying, "We want to control before planting" or the wheat or sorghum fella saying, "We want to control before the seed fills the head". The feral animal keeps going about its destructive business all year round, so we have to focus on the ecological or physical weakness of the animal and control them at that time. It doesn't always work, don't get me wrong. But that's the aim for me.

So I try to get at least one of the hooks into each landholder in the group, and a lot of that happens one-on-one or at smaller meetings to engage the key community leaders first. Then, I can work with them to develop a community engagement plan, possibly starting with a community meeting. At that meeting, I'm still trying to hook land managers, because they're not always all convinced and I'm trying to hook the people I haven't been exposed to yet. These meetings are crucial to get the community ownership, that's where the hard work is.

Land managers are really busy, so when you've got them in the room, you've got to be good. If you're not good, it's simple—they won't come back to your next meeting. Being good isn't just necessarily doing what they want—it means challenging them and helping them come away with something new they can implement or a joint way forward. At QMDC our funding sources are targeted at improving environmental outcomes. Often, a lot of the community is agriculturally focused. So I need to find a balance. In these meetings, we're addressing feral animals to improve both biodiversity and agricultural outcomes. If I were to call a meeting about protecting a rare mammal, I wouldn't get the same participation as if I called a meeting about saving 20% of a farmer's wheat crop or a large percentage of their lambs. But both of those meetings could conclude that the community needs to manage feral pigs and wild dogs, and both could have the same environmental outcome.

I try to take all the individual and organisational agendas out of meetings, field days and workshops. There is nearly always an individual in the community who is driving the agenda to control a certain species, and they will team up with the stakeholders that promote the same cause. This is dangerous, as it can often derail the focus of the larger community because they're just pushing their own barrow. It is important to address this, because without the entire community's ownership, coordinated control can't be sustained. Further, when there are only small grants and small buckets of money, there is a lot of competition amongst organisational stakeholders and industry groups to put focus on the specific pest relevant to them. In my opinion, it is just as important for organisational stakeholders to work together as it is for the community to work together. Sometimes, we need to practice what we preach.

So as soon as someone—whether an individual from the community or an organisational stakeholder—starts to push their own agenda in meetings, I put it to the side. In Australia, we have the "rabbit paddock", which is a box on the white board where we write topics to be discussed at a later time. An example would be, if we were talking about feral animal impact on lambing and a sheep producer raises wedge-tailed eagles as a major impact. These birds are protected so it is illegal to destroy them. So I would write it in the rabbit paddock and say, "We're on a roll here. I don't want to take away from your point at all, but we can talk about that after we've finished this subject". But then I cannot just put it to the side and ignore it, I have to come back to it. That's a John Gray trick, something he bashed into me. No matter how silly the topic may be.

And I have to be careful that $I'm$ not too focused on feral pigs and wild dogs either. If I sat down with any community group in our region and we had a discussion about their biggest pest animal problem, hands down it will be kangaroos. Feral pigs and wild dogs often aren't even on their radar. And then here we all are running around, talking about dogs and pigs. It's even harder if you're looking to improve biodiversity by reducing fox and feral cat numbers, because those species don't affect agriculture as much, at least in our area. From an environmental point of view, however, foxes and feral cats can have devastating impact on the native fauna. It is just so tough to balance biodiversity protection with agricultural-focused feral animal control.

I also need to accept that not all landholders see feral animal control as important; if it was, there would be more action at a community level. Not only that, some people actively work against it. Professional harvesters and wild pig hunters are just two examples. Not as obvious are some western Queensland cattle producers who will say, "The more dingoes in the landscape the better, because they keep macropods like kangaroos and wallabies down." If widespread community cohesion was strong, these cattle producers would control wild dogs, even though there is no impact on them, to look after their distant sheep and wool producing neighbours.

Even in communities where people prioritise feral animal control, the control method can be controversial. You can see that in the debate around exclusion fencing. Politically, exclusion fencing is meant to exclude wild dogs from enterprises that are impacted by them. But a lot of landholders are actually using it for macropod management. The kangaroo numbers are going to be so much higher on the properties outside the fence, and it's not going to take very long for those people to feel anger and hurt towards their neighbours who have decreased their kangaroo population by simply moving the problem. Putting up fences literally divides the community.

My end goal with education and engagement is to achieve coordinated feral animal control. But even if I've convinced people that control is in an important issue to address, and the community has agreed on which animals to target and when to target them—a feat in itself—I run into the problem that not all control methods can be effectively coordinated at a landscape scale. Sometimes, the methods that land managers want to use are just not effective, and it is my job to provide guidance as to what will work to reduce the population. It's an expert versus facilitator dilemma that's really tricky to balance.

I am making it sound almost impossible to get coordinated control to happen, but it's really not that grim. I think there's a lot of positivity in having more people working in feral animal control that are also interested in community engagement. I believe the overall fundamental problem is that the community, if left to their own devices, will never address feral animal control at a landscape scale. They need to be taken on a joint educational journey to discover that working together is the only way to achieve a real outcome on-ground. They need to know that individual ad hoc measures will have no serious impact on feral animal density.

My greatest challenge is how do I engage a community on a topic that they do not rate high on their individual agendas? This is a challenge I enjoy. First, I need to engage with the community about the issue, and then work with them to learn about how to address the feral animal issue at a landscape scale. This means not only bringing the information and learnings I have to the table but also listening, learning and incorporating the community's knowledge and experiences into the mix. When everyone has had input and agrees on the outcome, we wish to achieve as a community, the journey of feral animal management has begun.

Some people have a perception that people who work in feral animal management just shoot things or that we enjoy killing animals all the time. Especially fellas will say to me, "Oh you get to shoot pigs out of a helicopter. You're out there shooting them all the time". At the end of the day, what we're doing is what a lot of fellas pay money to do on the weekends. These fellas load their trucks up and drive from the city to go and shoot pigs. There's a real perception that my job's just fantastic and it's fun, and there's no pressure. Meanwhile, I'm out there collaring the pigs, trapping them, trialling new poisons and interacting with community members to get people on board, which is some of the more difficult work. It's not an easy job at all, but I'm sticking with it. I actually want to see a change on-ground and that is what keeps driving and pushing me harder. I want to progress. It doesn't matter to me what level I get to; I want to keep going. Every instance is a learning journey, for me and the community.

Chapter 13
Practitioner Profile (Greg Mifsud): "Pest Management Is All About People"

Abstract How can community engagement officers build relationships in new, unfamiliar locations? Greg Mifsud has gained this experience by engaging landholders across Australia. He has worked to build trust in communities where he didn't know anyone, and where he faced community hostility because of the history between key stakeholders. In dealing with these challenges, his genuine good intent and open communication has enabled him to connect with landholders. For Greg, it is equally important to change the behaviour of landholders as it is to change the way that government responds to issues. A lack of ongoing support and short timelines for initiatives can stand in the way of critical relationship-building. Greg acknowledges the discouragement that some community engagement professionals may feel at points in their career. Through both on-ground experience and training sessions, he has come to believe that a failed attempt at engagement is not necessarily a failure on the part of the engagement professional. To this end, he aims to reflect on his approaches, adjust them and implement new ones whenever possible.

I'm from Victoria originally; I spent my first 13 years in the outer suburbs of Melbourne. After that, my family shifted to regional Victoria. I've always had an interest in wildlife and animals. When I finished secondary school, I wanted to go to university and study Veterinary Science, mostly because I thought I had more of an affiliation with animals than with people. I did a degree in Biological Science, majoring in Zoology, Ecology and Animal Physiology.

I finished my degree at a time when most state government agencies were going through a rapid decrease in size and funding. Job prospects for university graduates in Victoria were especially limited; government agencies were getting rid of a lot of permanent staff, including many experienced wildlife ecologists, and employing them back as consultants. It was pretty hard to get into the mix if you were a new chum on the block.

Fortunately Dr. Pat Woolley, from La Trobe University, offered me a Master's degree looking at the ecology of Julia Creek dunnarts (*Sminthopsis douglasi*). The Julia Creek dunnart is a mouse-sized, nocturnal marsupial native to Queensland. I had completed an Honours year looking at these dunnarts in the laboratory but they

© Springer Nature Singapore Pte Ltd. 2019
T. M. Howard et al., *Community Pest Management in Practice*,
https://doi.org/10.1007/978-981-13-2742-1_13

had never been studied in the field. When I started in 1994 they were considered to be the second rarest mammal in the country. Although I had never intended to do postgraduate studies, I had always wanted to go to Queensland, to get out into the bush and see some of the country.

Dr. Woolley had a history of finding rare and threatened carnivorous marsupials. I knew there were plenty of foxes and cats up that way that would be impacting on native biodiversity, so I said, "Why don't we add foxes and cats to the project and look at their impacts on the dunnarts and other species?" I enjoy hunting, and I thought looking at the pest species might give me a chance to do a bit of shooting at the same time. Hunting is a bit of a cultural thing in our family, probably because we were from a Mediterranean background. I grew up hunting and fishing with my dad and uncles. It was very much a subsistence thing: you shoot it, you eat it.

My Master's project started my involvement with feral animal management and research. For a long time, that's all I had to write about because I didn't find a dunnart for nearly 18 months! I found a population of dunnarts on a property eventually, and I did a fair bit of 1080 baiting for cats and foxes to protect the dunnarts at that site from predation. 1080 is a poison commonly used in feral animal control and the toxin naturally occurs in many Australian native plants. As a consequence, most of our native fauna are relatively resistant to it but introduced predators such as cats and foxes are extremely susceptible. Through the research, I got a detailed understanding of the impacts that foxes and cats were having on native wildlife. I also saw how difficult it was to get money to study and conserve native species. I thought that the best thing I could do from a conservation perspective was to be involved in feral animal control, to try to conserve native species by managing feral animal populations.

During my research, I became a bit of a local identity "the dunnart man". I think that is where my community engagement experience started. I enjoy the company of people and I know that listening helps to build respect. Listening was a very important skill in Julia Creek because when I arrived I didn't know anyone. Most of the dunnart sightings were on private land. I was turning up at people's properties in a white vehicle with a University logo on the door, during a period when the government was buying up private properties for threatened species conservation. Talk about challenging!

In order to get their permission to have a look around, I had to build rapport, find out more about them and build some common ground. Showing a landholder that you are genuinely interested in them is really important. A lot of this was done at the local pub on a Friday night or various sporting or social events. My supervisor, Dr. Woolley, was a research academic and very focussed on the work. When we were in the field together and I was having a yarn with a landholder or a group of people, she would often be waiting at the car, tapping her fingers on the roof, muttering under her breath, "Come on. Let's get out of here". However, she later told me that without my skills to communicate with local people we would have got nothing out of that project.

After I completed my Master's, I spent a couple of years working with the Queensland Parks and Wildlife Service as a Technical Officer with the Wildlife Management Unit. The common threat to all of the native species I worked on in those two years was predation by foxes, cats and wild dogs. Then I found myself in another temporary role, this time with New South Wales National Parks and Wildlife Service (NSW NPWS) at Jindabyne. That was my first direct role in vertebrate pest management. My formal role was Pest Management Officer for Southern Kosciuszko National Park. There was a lot of political pressure at the time about the impact of wild dogs, so my primary focus was to get the wild dog management program up to speed and organise data collection so we could properly review the program and apply control in the areas where the impacts were greatest. There were two components to the program—baiting and trapping.

It was a difficult job. I was working across a number of organisations and trying to direct operational staff while having no immediate supervisory role, all the while trying not to tread on people's toes. On top of that, I had to work with the local landholders who had significant political influence over the program. The culture in Jindabyne was completely different from Julia Creek. In Julia Creek, and Queensland for that matter, you copped some light-hearted teasing about being a "parkie" but everyone got along and you still had a beer with the locals at the end of the day. When I got to Jindabyne, it was like being thrown to the wolves. I started on the wrong foot because I worked for the National Parks, and the local landholders had a long history of animosity with the agency. Kosciuszko National Park had been created through forced buy-backs of farming land. The National Park ("the Park") had put an end to high country cattle grazing and also restricted horse riding, which had a long tradition in the alpine area. So there was all this angst and anger within the community before I even mentioned that I was there to work on the wild dog issue. It took a lot of effort to get people to understand that I was there to help them, not to tell them what to do. I'm not sure I could have done that without the previous experience I had working with the landholders at Julia Creek, especially in terms of how to approach people.

I had to keep in mind that the angst they were feeling and their attitudes weren't personal. In the end, I was just very upfront. I would say, "Yes, I work for Parks; you can either work with me or tell me to get lost. It's your choice". Some of them would then say, "Well okay, we can sit down for a minute, I suppose". We were never best friends, but at least they respected me for being upfront. Another thing that helped was becoming part of the local community. For example, I used to play squash in town and go to the pub. Through that, I met a number of the local landholders and, slowly, word started to get around that I was honestly trying to help.

It didn't happen quickly. In the first week, I was given the wrong directions to a meeting with a group of landholders on one of their properties. I finally made it to the meeting place and was sitting at the kitchen table having what I thought was a reasonable discussion with a group of landholders about the wild dog issue, when one of them got up and said, "Oh, geez. Here we go again. They're going to spend $80,000 on another dickhead that doesn't know what he's on about". He then said to his wife, "Come on, we're out of here".

A week later, I rang and explained to him that I didn't know how the dogs were reaching his property to kill their sheep but I wanted to help. I couldn't see from the maps of his land how the dogs were accessing his property through the current control lines, so I said, "Do you mind if I come out there with you to have a look?" There was stunned silence for about two minutes. I said, "You still there?" And he replied, "Yeah, yeah, still here". I said, "Would afternoon tea today at 3:30 be okay?" He said, "Yeah, alright". I got there and we jumped in his truck on our way out to the paddocks, when he went very quiet. I said to him, "You're a bit quiet. Is there an issue?" He just shook his head, looked at me and said, "No mate, you're just the first guy in 50 years that's ever come out to check what's going on at my place to try and help me solve my problem". It was always a tenuous relationship with that property-owner but I think I earned his respect that day.

His brother also owned a property in the area and was more of a community leader, one of those old-timers that the rest of the community looked to for direction. I could never really gauge how he was going to react in a public meeting. Sometimes, he would say nothing and at other times he would really whip the landholders into a frenzy over the wild dog issue. I thought that if I could gain his trust, then perhaps I could get more traction. A good word from this guy would be the catalyst to get the community working with me and moving forward.

I was driving back from Byadbo Wilderness area in the southern end of the Park one afternoon when I saw him walking through the paddock back towards the road. I was in a National Parks vehicle, and I slowed down and waved to him, indicating I was going to stop and say g'day. Then I realised, he had a gun in his hand! It would have been rude not to stop at that stage, so I pulled over, all the while thinking, "You're an idiot. This isn't going to end well". Anyway, we got to chatting and I asked him how his day was going. He said, "Not good—I'm just trying to sight in this bloody rifle, I can't get it to shoot straight". I asked him what calibre of rifle it was and he replied in a rather surprised tone, "Oh, it's a .243 Winchester". I said, "That's what I've got". So we found some common ground and talked about shooting and rifles for a little while before I got up the courage to ask about the wild dog issue. I said, "I'm just trying to get my head around what's going on down here with the wild dogs. Do you mind if I sit down and pick your brain?"

I reckon we sat there for nearly an hour-and-a-half, on a log in a paddock, talking about the history of the region and how things had gotten so bad locally. We didn't end up talking much about dogs really—it was just an opportunity to learn about the culture of the community. He had his public face that he had to maintain, and I always try to keep that in mind—some people react differently in public than they do in private. His positive influence on the rest of the community in terms of working with me and the wild dog program was immense from that time on.

During my time in Jindabyne, I worked with Peter Fleming from the New South Wales Department of Primary Industries (NSW DPI) and Rob Hunt from NSW NPWS to develop nil-tenure management plans for all of the wild dog management groups surrounding Southern Kosciuszko National Park. Nil-tenure is a planning process where local communities work with government land managers to address pest issues across all land tenures in the area, rather than trying to lay blame for where the dogs reside. The process starts by looking at landscape scale maps without the property boundary lines. Getting rid of the property boundaries means you can focus on the landscape and how the wild dogs travel to access livestock populations.

By focussing on the geographic aspect of the problem, it's possible to work with the community to deliver control methods in a more strategic manner. The aim is to expose as many of the pests to the control method as possible.

The nil-tenure process relies on bringing everyone into the room and giving them an equal say in the decision-making process, something that had never happened in that region in the past. For the first time, landholders were invited to be part of the process and have a say in the decisions that were being made. The NSW NPWS still had limitations on what could or couldn't be done in regards to dog control inside the Park due to legislation, but once the landholders were informed of the reasons why, and felt they were part of the process, the animosity was taken out of the issue. Stock losses declined as people started working together. Baiting programs were delivered on private land in conjunction with baiting programs in the Park, and we had wild dog controllers available to deal with stock attacks as they arose. The process worked so well and the impact on livestock was significantly reduced. The animosity and anger associated with wild dog management was replaced with healthy working relationships between the landholders and NSW NPWS.

In fact, the nil-tenure process worked so well that people stopped complaining about wild dogs; complaints to the state minister ceased and as a consequence, it was deemed that a Pest Management Officer was no longer required and my contract wasn't renewed! That took a big toll on me personally, especially after the amount of work I'd done with people in the area and the camaraderie I'd built up. I felt like part of a team, like part of a very close community. I had gone into the work thinking that if I did a good job, I would be rewarded with a permanent position.

It also took a big toll on me professionally because it was a very challenging role. There's a lot of effort involved in working with small communities. Some people embrace you with open arms because you're willing to help. Others are very sceptical because they've been burnt by the government before. It's even more difficult when you work on an issue that is emotionally and politically charged like wild dogs.

We decided to move back to Queensland and I started applying for permanent jobs. I was offered one with the Terrestrial Planning Unit at Queensland Parks and Wildlife Service (QPWS). I was there for 2 years working on the southeast Queensland forest agreement transfers. Large parts of the state forestry estate were being handed over to QPWS as conservation estate of various tenures. My role was to work with stakeholders that had stock-grazing permits or leases for these forest blocks and inform them that those permits were not going to be renewed. We ran a "grazing hardship" program that gave landholders the opportunity to show how much of their income was based on the pre-existing permit or lease. If the landholder could demonstrate that the lease or permit they owned contributed to at least half of their on-farm income, then they could apply to have their lease extended for a period of time in order to shift their business interests and move away from their reliance on the lease or permit area. It was a very fair process—only four landholders required an extension of their lease in the end. I was proud of that outcome.

I then went on to another permanent position in local government with the Gold Coast City Council Conservation Management Unit. I was responsible for managing their conservation areas, mostly blocks of land left over from housing developments that no one else wanted. After a while, the opportunity came up to become the Council's Pest Management Officer. There was a bit of pressure around this position because the Council hadn't finished their local government pest management plan, which was a regulatory requirement. The Council thought the state government had failed to undertake adequate control of pests on state-owned land and were still angry that the state had divulged the responsibility for management of pest plants and animals to local government when they changed the legislation in 2002.

The other problem was within the Council itself; there were a lot of different management units involved in pest management—weeds, biosecurity and conservation—and none of them talked to each other. My role was to work across the different directorates to develop some sort of consistent approach to pest management. That was difficult because many of the managers viewed my position as a threat to their department. Everyone closed their doors, thinking I was there to poach their staff. So I started going down and chatting to the managers in the different divisions, having lunch with them, playing darts in the lunchroom and building relationships. Once I proved that I wasn't going to steal their staff and that my intent was genuine, things were good. We built joint management programs and improved communication between each section of Council to prevent weed-spread throughout the Council area.

In 2007, the role of the National Wild Dog Facilitator was advertised and I successfully applied. Through my previous roles in pest management, I felt I knew the players in the wild dog space; and even though there was a lot more travel involved, the job got my young family back to Toowoomba, where my wife comes from.

It's a national role, funded by a partnership between industry bodies and the federal government and hosted by the Queensland Department of Agriculture, Fisheries and Forestry. There is also work at the state and local level. The aim of the role is to roll out the nil-tenure approach as best practice across the country. At the start, I had to sell the nil-tenure approach to various state governments, and establish and work with local community groups. It was easier working with the state agency people because I knew most of them. When I went out to the local areas, I had no real knowledge of the people in those communities—apart from the ones I'd already worked in—but the locals picked up quickly that I had experience working with communities and had pest management knowledge. I also think that working for the IACRC, an organisation that was independent of government, gave me a bit more credibility.

In 2008, I established the National Wild Dog Management Advisory Group (NWDMAG) to advise me in my role as National Facilitator. This group was unique because it pulled together farmers and industry from all over the country. At one stage there were actually more landholder representatives on the committee than government. Working with the NWDMAG was a real catalyst for the wild dog program. I took the members outside their local area to see the wild dog issue across the country. That got them thinking about wild dog management on a national scale and exposed them to how other people were dealing with the issue. In many instances, they took that experience back to their relevant states and communities and were catalysts for change amongst their local communities, but also across government and the broader state level.

Through the NWDMAG I've meet key people who are strong within their industries and who can support me when things get tough. It can be hard to see the light at the end of the tunnel sometimes in this job, so it's great having mentors to bounce ideas off, or to have a beer and vent your spleen when things get really frustrating. These facilitation roles are difficult because it can take a lot of time and effort before any change occurs. In this job, you very rarely get closure, and it's often hard to see whether you are making a difference. I think that coming to grips with that, and having a network to provide support is important; I work to provide this kind of supervision and mentoring to other wild dog coordinators around the country.

Now, my focus is primarily at the national level. I look at big picture things like training and policy development, implementing the objectives of the National Wild Dog Action Plan and working at a higher strategic level. It has been hard adjusting. I miss being in the field at times because I like being at the coalface—sitting back is always hard for me. I love getting to know people, getting to know the issue and learning about communities. There's a lot of policy based on hearsay, but when you get out there and talk to people, you get a real understanding of the issues on the ground. You can't truly help communities move forward until you've got a proper understanding of what's going on at their level.

The other side of those interactions is that you become attached and develop friendships with people in those communities.

Successful long-term relationships with community groups are built on trust. I am still actively engaged with community groups I commenced working with 8 years ago. It's hard not to get emotionally involved. The people I work with often end up being my mates. As my role has evolved, there seems to be less time to keep in touch and see how they are going. With the droughts and everything else landholders and farmers have to deal with, it can be a lonely place for them out there.

Community engagement in the pest management space is about achieving a cultural change and shifts in management behaviour. My role is to bring stakeholders to the table and create the space where they feel that everyone has a fair say and equal opportunity in the decision-making process. In order to bring the community along for the ride, I need to take the time to understand the community and the dynamics at play. It's hard to try and generate cultural and behavioural change when you don't live in that community or have a pre-existing relationship with the stakeholders.

It's not just the landholders who struggle with change. There's a behaviour change needed within government too. What really got to me at the beginning of this job was the government staff who would say, "That's not how we've done it in the past, so why would we do it like that now?" I couldn't really get any commitment from anybody to work with these groups. Looking back, some of the plans I developed with stakeholders fell over because they did not get the ongoing support that government promised at the start . Using a nil-tenure approach to pest management planning is such a big shift for some landholders and, if someone isn't there to provide support and remind them why they're going down that path, they quickly revert to what they used to do.

However, just because a community engagement program doesn't deliver the expected result or outcome doesn't necessarily mean it is a failure. When I first started, I thought that if the group didn't get organised that people would frown upon me because I hadn't achieved my purpose. I set numerical targets each year for the number of groups I was going to work with or write plans for. I don't do that anymore, because if one group doesn't come on board then I don't reach my milestone, and it is no slight on my capacity if a group or person chooses not to participate. Overall, there have been some big wins in this space. I think it's important to reflect on these achievements rather than always focussing on perceived failures.

More recently, I have undertaken some community engagement training with staff from Penn State University. I think it's a good idea to do community engagement training. Community engagement has only recently been considered as a professional skill here in Australia. When I started in this role, training consisted of being given the directions to the town hall or property where you were to hold a meeting with landholders—and heaven forbid if the directions were dodgy! Knowing more about the different processes and how to deal with the different characters helps me plan a process to suit a group. As a facilitator, the only thing I can control is the process. At the end of the day, the landholders must do the invasive species control.

Many landholders do a lot more work on wild dog management and in community groups than they are given credit for. I feel frustrated when I see landholders being burnt out by a lack of government support for community participation. Landholders that volunteer their time to support community-led action are limited in their capacity. They can't go out and force people to do dog control. They have no jurisdiction to get the contact details of people or create maps. That support has to be provided by someone or somewhere else. Society can't expect the landholders to undertake roles or deliver programs that require involvement from statutory authorities. At the end of the day, they're landholders trying to run a business, keep food on the table, raise a family and look after their land—on top of volunteering to organise the local pest management program. That's where I get concerned about the whole community-led approach. Landholders can only take things so far and those processes are going to fail if they don't get the support they need to keep going.

The current government trend of reducing regional staff and moving away from one-on-one landholder interaction has resulted in a large proportion of landholders not getting the information they need to understand feral animal control or that they have a requirement to participate in control programs under state legislation. Ultimately, it's necessary to have pest species officers out on the ground talking to people.

I don't assume that people who do not participate in control programs have made a conscious decision about that. Many landholders don't realise they have a legal obligation to control wild dogs until someone visits and speaks to them about it. Some do know they have an obligation, and think that carrying a gun in the car and shooting the odd dog is sufficient. There are also misconceptions about 1080, like that it kills any native animal that eats it or it threatens the safety of working dogs. Once they understand that 1080 is very species-specific—especially at the low concentrations used for introduced predators—and show them how to reduce the risk to their working dogs, they will often get on board with control. Unfortunately, especially in rural areas, government budgets are limited, staff resources are stretched and the distances between places make compliance processes very difficult to carry out.

The nil-tenure approach is not always applicable either, because the community capacity might not be there. In the past, particularly in rural and remote areas of the country, everyone was part of a community. They felt a civic duty to participate but these communities have changed so much now. There's still the odd family place, but now there's lots of people managing properties from outside the area, a lot of absentee owners. I'm reassessing my approach in some places to consider options like the cluster-fence model, in which a group of like-minded landholders put up a fence around their properties to look after themselves, because they can't rely on their neighbour to manage pest animal numbers.

I'm also reassessing how I communicate with people. People often think if they are good talkers that they are good communicators, but I don't know if that's the case. One thing I've learnt through community engagement training is that communication is telling someone something, while engagement is listening. I'm trying to work on that because from my perspective, pest management is all about people. Controlling the animals is the easy bit.

Chapter 14
Practitioner Profile (Mike Reid): "It's not Perfect, It's Complex—And That's Ok"

Abstract Navigating community and government power dynamics can pose a challenge to engagement practitioners. In this profile, Michael Reid describes how he learned to not just navigate, but meaningfully reshape, power dynamics in such a way that levels the playing field between landholders and policy-makers. The son of a South African immigrant, Mike felt drawn from a young age to learn about power and to work for social justice. At university, he combined an Applied Science degree in Agriculture with a degree in Sociology. Throughout his career, he has sought to apply lessons from both fields—framing issues to highlight the social and political dimensions as much as the scientific ones. By bringing landholders to the seat of government and government officials out to farms, he seeks to build common understanding and create opportunities for everyone to have a voice and tell their own story. Mike believes a collaborative approach to learning can allow for a fundamental reimagining of ecosystems and political systems alike.

I grew up in Albury–Wodonga, on the Victorian side of the border. My mum is from a cropping place called Berrigan and my dad is originally from South Africa. They represent two very distinct cultures. My mother's side of the family used to have a few farms around Berrigan, near where I grew up. I can remember, as a young guy, sitting on a combine harvester with my grandfather and harvesting the wheat. Although my uncles did really well out of farming, an economy of scale swept through the area and those smaller family farms just weren't viable anymore; it's kind of sad. They were all very caring and gentle people.

My father's family had more of an old school, South African mentality, with a strong sense of stoicism and fortitude—a stiff upper lip. You worked hard and pushed through, because that was just what you did and who you were. Maybe that's part of an immigrant mentality. My grandfather on my father's side moved to Australia from South Africa when I was in my early teens. He had a passion for nature; we connected around being outside, gardening and growing plants. He used to bring seeds from South Africa when he'd go back every year and we'd grow maize and other different things.

I was always interested in agriculture and botany. I remember, in high school, reading a book about careers and there was a picture of a guy standing in the paddock with a computer. I thought that connecting agriculture and technology would be the way that I could help people. Growing food is such an important part of people's lives.

Some of the family on my father's side started immigrating to Australia in the late 1970s, which was at the height of the apartheid era in South Africa. When dad first arrived in Albury–Wodonga, the *Border Morning Mail*—the local newspaper—put a story in the paper about a white South African moving to the district. It was a big deal at the time, and the attention wasn't always positive. As I was growing up, my father taught me some Zulu words, which I always thought were Australian slang until I went to school.

I wanted to learn more about the culture. I spent a lot of time in the library reading books about South Africa and political prisoners such as Steve Biko; I would think about the injustices happening there, and what it meant to have that as part of my heritage. Inequality is the issue that makes me most angry. When I was 14, my dad took me over to see where he grew up, and to experience the culture for myself. It was a special time for me. We went to Cape Town and saw Robin Island where Nelson Mandela had been jailed. It's a pretty complex place, South Africa.

The story of Nelson Mandela always fascinated me. He was President of South Africa when I went there with my dad. Mandela spent half his life in prison unjustly but when he got out he said, "Let's all say sorry to each other". What he did for the nation remains a big inspiration for me. I've been to South Africa three–four times since then. Though it's rich by the living standards of other African countries, there's still a lot of poverty. I'm passionate about social justice, and understanding how power is exerted has influenced a lot of my thinking. When certain groups have more power than other groups, and that affects delivering a good outcome for everybody, it motivates me to make a change.

After high school, I studied a Bachelor of Applied Science (BAS) in Agriculture at Charles Sturt University in Wagga Wagga, New South Wales. I was doing my degree by correspondence and loved all the subjects but about halfway through, I got to questioning, "How would I be able to help farmers with this information?" I wanted more than just a narrow technical focus. So I started a Bachelor of Arts (BA) in Sociology at the Wodonga campus of La Trobe University.

I pursued both degrees concurrently, through the two different institutions. The study helped me look at the bigger issues that were happening in rural communities in Australia. I thought that I could take the technical knowledge from the BAS and potentially go to Africa and improve crop yields. Then the BA opened my mind to the naivety of that thinking. I started asking: "What is progress? What is development? What does it mean for rural communities? What would that mean overseas?"

A pivotal book which shaped my thinking was *The Violence of the Green Revolution,* by Shiva (1991). The Green Revolution was a technical, political and ideological revolution that started with the advent of hybrid wheat and the use of synthetic fertilizers and pesticides in India. It improved production, but the economies of scale led to an intensification of inequality. Rural suicide rates just went through the roof.

Win-win situations aren't always possible when it comes to agricultural development, and I think that's the benefit of critical thinking—to look at the implications and consequences. Complexity is there, but it doesn't always get recognized.

This influenced my career direction. In 2002, I managed to hustle my way into a position as a Research Scientist with an extension-based group at the Rutherglen Research Centre, which is part of the Agriculture Victoria Department (the Department). I focused on how to improve engagement of small lifestyle farmers across the state. Having both degrees was a big help because I understood social science methods, but also had knowledge of agricultural science, which gave me some credibility with farmers. I mapped demographics like farm sizes and incomes, and conducted interviews and focus groups.

Around this time, Albury–Wodonga had started to see an influx of African refugees coming in from the Congo and from Sudan, and also some skilled migrants from Nigeria. In 2007, they organised a community meeting and asked if I would facilitate it. I rocked up to a local hall and there I was—the white guy—with 70 refugees and skilled migrants. It was a scary time.

I posed a question to the group, "What are some of the issues that you're dealing with in regional Victoria, as an African?" They talked about a number of issues, and from there I said, "So what do you want to do about it?" And they said, "We'd like to form a group". So we organised a subcommittee to form a group, which eventually became an incorporated body called the African Union Albury–Wodonga, to represent African needs across the region and to run African cultural events. We were organising our next town meeting when a member of the group phoned to nominate me for Secretary. I said, "No, the constitution states you must be African". But because my father was from South Africa, that allowed me to be nominated. So I became the Secretary for the African Union Albury–Wodonga, which was a very strange and rewarding experience.

In this role, I helped the group navigate regional society in Australia. We organised local events and they would ask, "Look, Mike, we are unsure what to do. Who do we contact?" They wanted to make sure that while they were representing their cultures to a regional community, they were also getting the Australian customs and traditions right. That was probably my first step into community action. I would do things differently with respect to cultural sensitivity if I had the chance now; I would have tried harder to acknowledge and address the diversity across the group. But it went well and it served a purpose and we got some good runs on the board from that. The community has continued to grow. At the first meeting we had 70 people, and now it's probably about 150.

During that time I took on a new role within the Department, as a Senior Pest Management Officer. This role was a coming of age for me in terms of community engagement; in this capacity, I served as the Executive Officer (EO) of the Victorian Blackberry Taskforce (VBT). The EO had two roles. The first was to provide executive support, which included preparing meeting agendas, organising meetings, overseeing budgets and strategy, maintaining the website and bridging the relationship between government and community partners. And the second was to help establish community partnership groups.

Previously the government approach to controlling blackberry had been to go out and spray the blackberry themselves. Then they moved to a compliance model, in which they'd send people out to direct farmers to control the weed. In the worst-case scenarios, the farmers would be issued fines or taken to court. Eventually, the government pulled out their Compliance Officers in many areas and said, "The writing's on the wall. We can no longer financially support compliance and it's time for a community-led approach". This is how the VBT was developed. When I started as EO, the group had only met once or twice. The Taskforce—which initially consisted of two people—was lacking direction and connection to on-ground programs. This changed when Lyn Coulston, a community leader from northeast Victoria, got involved.

Lyn was the Mayor of Towong Shire, which incorporates most of the Upper Murray region. At the time, they had a terrible blackberry problem. Blackberry was really well suited to that steep country, and was growing for kilometres along the fence lines. In 2005, when Lyn learned that the government was pulling out of the compliance model, she put together a community meeting which initiated the North East Blackberry Action Group (NEBAG). She got someone to facilitate, and the group developed a one-page action plan for how they wanted to address the issue. They still use this plan today. They wanted to employ an external person to do extension and property management plans with landholders. They knew that they had to get somebody in to assist, but not a government person. Following this meeting, Lyn approached the state government for support with the plan; because of her interest in the issue, she was invited to join the VBT. At the first meeting she attended, the other two members nominated her to be the Chairperson.

The VBT had no budget, so Lyn asked for a meeting with the Minister of Agriculture. Based on that meeting, they were able to secure funding for the VBT and NEBAG to run their program. The group is now celebrating their 10-year anniversary, and that was the only grant they ever received from the state government.

The first Project Officer that the group hired, Damien Wall, was able to build good relationships with people. Three things made Damien effective: he wasn't from the government, so he had trust; he wasn't from the local area, so he didn't carry local politics; and he had some good technical knowledge about how to control blackberry. He'd go out with people and map the blackberry on their property with GPS, and he would bring public land managers from adjacent properties into meetings to coordinate efforts. Then he'd create a property management plan, and the landholder would sign it—these aren't legally binding, but there's community peer pressure.

NEBAG was the basis for the VBT model of grassroots, community-led action. If you went to one of the NEBAG meetings, you'd see the local community members there and you'd see the public land managers there. They'd have a pizza and they'd have a cup of tea, and then they'd get down to business and reflect on what they'd been doing. Everybody would have their say. For some people, come to those meetings and feeling like they're contributing is a good outlet. In the middle of the mountains, in a country hall, people are making decisions and taking control.

One of the things the VBT identified early on is that—although we know how to control blackberry from a scientific perspective—if we started looking at blackberry as a people issue, it opened up a whole set of different solutions. And that was our mantra: "We're going to define this as a people issue: blackberry control is more than science". This story is captured in a short documentary (North East Blackberry Action Group 2009). We started to document that story in a Capacity Building Toolkit, based on the northeast experience, which community groups could use to help organise their programs.

Lyn and I would attend community meetings and present the government policy. Lyn would tell her story about how NEBAG addressed the issue. She wouldn't sugar coat it—she'd just tell it how it was. Then we'd facilitate a session and ask "What do you think about this approach?" People grappled with the issue and how their community could deal with it. If they wanted to run a similar program, they could submit an application to the VBT for $20,000 over three years to implement a community-led approach. This money was for group development, not on-ground works. The federal government was already providing subsidies for spraying. But to build long-term capacity, it's important to move people away from that subsidy mentality. It's the difference between a hand-up and handout.

Since then, the VBT has grown into an important institution that has lasted three changes of government, and four departmental name changes. It has representatives from all across the state, and it has replicated this community model—with varying success—across 14 groups. Each one of the projects the VBT has funded looks different and has different results because the 14 communities are different, the issues that they're dealing with are different, their capacity is different and their views on the role of government are different. The goal is not to create a cookie-cutter approach—it's to meet those different needs.

The VBT has continued to gain recognition within government. Twice the Executive Directors flew into Albury and I took them up into the mountains, to meet community members at their level, in the local hall, and then to go out to farms and explain the program that way. It is important for key decision-makers to see how things are on the ground. When you are looking at a hill of blackberry, it's clear that there is no market-based solution. We produced an annual report each year about what the community was delivering in terms of blackberry management, and we sent that with a bottle of blackberry jam to the Minister of Agriculture. The government offices are in Melbourne, and that's where we'd have our quarterly meetings. I enjoyed bringing farmers and community members into that corporate environment to participate in the decision-making process and merging those two cultures together. Democracy is about bringing people into that space, and overcoming the tyranny of distance.

For me, facilitation means stepping back, letting those complex stories be told, connecting those people up and then seeing what happens. I was lucky to work with Lyn, and learned a lot from her approach. Lyn can hold her own with the elected representatives and the government executives in Melbourne. When she sits down with an elected minister, she can represent the VBT's needs while also speaking the strategic and tactical language of politics. To be able to carry both those languages and speak both those languages is a unique skill.

It wasn't always smooth sailing. There were tensions. In the early days with the VBT, I was passionate and didn't understand some of the internal departmental politics and power dynamics. I'm a very emotional person. Although we had support at a high level in government, a lot of the compliance staff would say, "No, you're not coming into this community". For me, it was like a rugby match. I had the ball, and a line of compliance people to push through. I felt isolated in the Department, pushing this approach, and that was hard.

I was in the EO role with the VBT for about 7 or 8 years, and by the end I was emotionally spent. I needed a break, so I took a year sabbatical. During that year, I went travelling, stopping in South Africa, Mozambique and then the United States to stay with my girlfriend, who has since become my wife. I stopped in Mozambique because, prior to leaving, I had done a Masters of International and Community Development by correspondence through Deakin University in Geelong, which included a small thesis exploring participation and its implications for food security and natural resource management in Mozambique. In the US, I worked for a while as the world's worst waiter in Las Vegas, and that was tough. I missed my family and was working a job I didn't like in a city I didn't like. Then, a contact in Victoria, Professor Tim Reeves—who had been the Director of the Rutherglen Centre before my time and had done a lot of international work—gave me some contacts to follow up at US universities.

I went to Stanford University, in California, and met a gentleman called Wally Falcon, who was a Professor of International Agricultural Policy and Economics there. We had a good conversation over the course of day and then Wally invited me to submit an application to become a visiting fellow. I spent about 4 or 5 months at the Stanford Centre for Food Security and Environment, auditing one of their courses on global food security and agriculture. During that time, I was exploring some of the wider supply and demand drivers of international agriculture and development.

That sparked a passion in me and, when I returned to Australia in early 2012, I enrolled in a Ph.D. program at Melbourne University. My subject is the sustainable intensification of agriculture. I never wanted a traditional career in agriculture. I'm interested in looking at the intersection of community and science. My goal is to understand the perceptions within different institutions of sustainable intensification. Sustainable intensification is an aspirational policy concept for achieving food security, which aims to produce more from the agricultural land that we've got in a way that is more environmentally and socially viable in the long term. How to put this into practice is very contested.

At the same time that I started the Ph.D., I took on a Senior Policy Analyst role in Melbourne for the Victorian Government's Agriculture Policy Unit. In that role, I was responding to requests from community, industry and local government stakeholders, and providing advice to the Minister for Agriculture. It helped me understand more about how the political system and legislative process work, how political decisions are made, how different groups are consulted and how our parliamentary process works.

It was there that I met a Lisa Adams, who was at the time the National Rabbit Facilitator with the Invasive Animals Cooperative Research Centre (IACRC) (see

Chap. 4). Because of the work I had done with the VBT, Lisa involved me in some systems mapping that she was doing around rabbit management and, when she took another position within the government, I was offered the opportunity to fill her role. That was a hard decision, because my wife had a baby on the way and the policy role was a permanent position with job security that the rabbit position didn't have. But I wanted to get back into engagement and I wanted to work with Lisa. So I took on the rabbit position in November 2014.

Lisa is very ethical in her practice and her leadership. To create the systems map, she collected stories from a range of different people and ran a workshop based on those stories. This became the basis for the project strategy. That was a big learning for me. Whereas the VBT had been more grassroots, the rabbit work was very strategic, and there was a lot more time spent building consensus across the various stakeholders within the rabbit management system.

A community group called the Victoria Rabbit Action Network (VRAN) was established to oversee the implementation of that strategy. VRAN has four community members and four government members, of which I was initially one. I provided secretarial support, whereas the other government members were more active participants. In 2015, we ran a course called the Rabbit Leadership Program, with 35 people from all over Victoria who came to learn about integrated control of rabbits, their legislative responsibilities, and community engagement. From that course, we developed a Learning Network of about 22–25 participants.

I had to take some big leaps of faith in my role as Facilitator of that process because I'd never been involved in a learning network before. It has to be self-organising and completely voluntary for people to come along and continue that conversation. The longer the learning network goes on, the smaller my role is becoming, which is fascinating. For example, recently the learning network came together at Ned's Corner, which at 30,000 ha is Victoria's largest property. Rabbits have had a huge impact on that landscape. The group spent eight hours out in the paddock, going to different sites and talking through issues. At the end of the day, we went down to the Murray River. My energy was flagging but it was important that we did an evaluation. I told everyone to grab a chair while I went to the bathroom, and when I came back everyone was sitting around in a circle. I thought, "Well, I'm not going to disrupt that", and we sat in this circle for an hour-and-a-half by the river, reflecting on what we did that day and what it meant. I barely had a role. It was a powerful moment for those people involved. From a facilitation perspective, I didn't know that it would happen—that people would be able to connect in that way. But the relationships were there.

After that meeting, I didn't write up notes but sent out an email with some thoughts. One of the guys—a rabbit management contractor with a grain harvesting business raised the question, "What's our purpose?" and I said, "That's a question I need to think about a little bit more". Then he responded in another email that said: "Hey, I just want to tell you what our purpose is from my perspective; this is a good idea of what we can do next; and this is what I've learned from being part of the Learning Network". At the start of the field day he had been up the back of the group, with his hands in his pockets, kicking the dust—and by the end he was bringing his voice to the table. Witnessing those changes has been really moving for me. I can go home at the end of the day and feel like I've made a real difference.

People ask, "Well what does the Learning Network mean for managing rabbits? Is it just talk?" From what I can see, the Learning Network is a transformative process. Some people go back to their organisations and change the way that they work with the community to establish rabbit action groups. Others go back to their communities, access funding, take leadership roles and organise events.

Right now I'm telling you this story, but in Victoria, it's the people who are telling it. There is always a power and a politic in constructing stories. With the VBT and the VRAN, there are a lot of good things coming out of both those groups, but there are also some tensions and areas which are problematic. The best people to tell this story are the people involved, and that means a "warts and all" approach: "The program is complex—it's not perfect—and that's ok".

The VRAN and Learning Network have opened up the space for participants to do the talking. That's the best promotion the program can have. From my perspective, it's redefining local and state politics by reinvesting in people. This program has been contested, but I'm learning that some conflict is ok. For example, there's pressure to try and replicate the program: "Okay, it's worked in Victoria. Now do the same thing in other states and territories". But it took a year to nut out this system, to get it working in this particular place, with this particular community and I'm not going to compromise by rushing to roll it out elsewhere. These things take time.

There are some strong stereotypes about farmers. Some people would describe them as not wanting to say a lot, or not wanting to participate. But the farmers I've worked with are committed to land management and making positive changes in their community, and they've got the high-quality leadership skills to do so. For example, the Chair of VRAN is very professional and skilled. I don't want our relationship to be that we're going to talk about farming and then I'm going to use that to influence him. I've done that kind of engagement where I've built up a rapport and use that to be persuasive. Instead, I go to him for advice because he's got good political skills to contribute. That's a different type of engagement, when you strip away those stereotypes and connect around skills and passion.

In community engagement, it's important to create safe spaces to fail and be able to talk about that failure. Sometimes you might have to make hard decisions or compromise on one of your values in a meeting. I want to be really honest and transparent about that. Government as an institution doesn't do that very well and that's why governments can really struggle to innovate, while there's so much innovation at the community level and in the community system.

Thinking back to the group managing blackberry in northeast Victoria, they've been going about 12 years and they've done a lot of work. If you asked them, "Have you seen a reduction in blackberry?" they might tell you that they've seen a little bit of a reduction. You might then think, "Well, that program hasn't been very successful". But just imagine if this group hadn't intervened, what the landscape would look like now, and what relationships in the community would look like now if there wasn't this intervention. There would have been a withdrawal of government services and nothing to take their place. Because there's been a blackberry action group, it's given the community a common cause to rally around—something they all support. Invasive plants and animals haven't changed—maybe a little bit with technology and a few other things—but not a lot. What's actually changed is the recognition that these are complex issues.

I think there is a great opportunity to shift the narrative about what "success" looks like; how we manage weeds and pests in Australia; and what community means in that space. The story can move beyond a focus on just eliminating weeds and pests to talk about public benefit. Through the rabbit project, we've been able to attract $4,300,000 of federal government investment in agriculture to Victoria, to extend the program to work with other communities on invasive species management beyond rabbits. There is funding support for community-led action and learning networks, and strong governance to allow the community to deliberate on how that funding is spent. I recently moved into the role overseeing this funding, as Program Manager for Established Invasive Species.

As I move forward, my engagement practice is going to be more deliberate in terms of how I go about working with people. Just paying attention and actively listening to what someone is saying can be a political move, and a deliberate move. There is a sense of the political worth of these collaborative approaches, but there are a lot of different voices across the table. My personal challenge is to make those engagement processes ethical and respectful, and to foster a spirit of collaboration.

References

North-East Blackberry Action Group (2009) Blackberry control is more than science. YouTube video file. https://www.youtube.com/watch?v=pIese4Csb-8. Accessed 12 December 2017.

Shiva, V. (1991). *The violence of the green revolution: Third world agriculture, ecology, and politics books*. London: Zed.

Chapter 15
Practitioner Profile (Harley West): "Whatever They Say, I Treat It as a Serious Question"

Abstract Some people's careers follow a straight line while others' are full of twists and turns; Harley West was in the latter category. After 15 years as an organic farmer, he fell into an engagement role. As an engagement practitioner, West was determined to make sure that all stakeholders were at the table and that their voices were taken seriously. He sought to ensure that everyone who attended a meeting walked away feeling as though they were involved in the decision-making process, whether they were for or against it. He observed that when people walked away from a meeting feeling unheard or disappointed, they had the potential to become "silent destructors" who would do anything to sabotage plans moving forward. Engaging people in such a way that everyone feels involved can be daunting. In addition to interacting with farmers during workshops, Harley also held one-on-one meetings to understand people's perspectives, noting that these collective and individual interactions complemented each other. Though he did not set out to become involved in community engagement, Harley ultimately found himself in a position that seemed as though it was made for him.

I guess if you wanted to typify all the major things in my life, you'd have to say that they were accidents. I've never had ambition—I've never looked for things—but I've always had plenty of work to do and I've always had reasonable people around me.

I met my wife when I was in my late-20s. We got married, and we lived in Ireland for a year. While I was over there, I got really interested in herbs. We came back to Australia and I started working in an organic fruits and veggies and dry-goods outlet. I was the fruits and veggies coordinator, so I'd talk to farmers on the phone every Friday, and then meet them when they dropped off their produce. We'd stand around the car and chat, and they seemed to be quite nice people—some of them just love to talk. I started developing a nice appreciation for farmers.

I also started a natural therapy course when we came back to Australia. As I was coming to the end of that course, my wife and I found the farm in Stanthorpe, Queensland, where we live now. It was completely accidental the way it happened; my wife's sister found the place and told us about it. A retired farmer had set it up. He was a bit of an experimentalist and had tried out a whole lot of interesting ideas on the farm. For example, he'd done a lot of close planting, where you plant grape vines close together. This was not done at all by the vignerons, or winemakers, in the Granite Belt area of Queensland at the time, but it's now quite a common practice in the region thanks to him. He'd also planted a mixture of crops; he had apples and peaches, nectarines and different kinds of grapes.

We basically went on a learning curve, teaching ourselves to grow; just learning by doing, reading, trying to talk to people. Because we came from an urban organic background, we weren't going to be using any chemicals. Everyone around us said, "Well, you can't grow anything if you don't use chemicals". But we did grow things, and we became growers. Then we went into small crops, so we had plums, zucchinis, tomatoes, garlic and snow peas.

I did most of the farming. I learned all about tractors and pumps and irrigation. It was tough, digging a hole and finding the two-inch irrigation pipe as it spurts out at you—bugger! We never made money, but we never lost any either; we were always able to buy new equipment each year. It was good, enjoyable. We farmed from about 1990 or 1991 to about 2005 or 2006, around 15 years. We saw it all—the hail storms, bush fires, frosts—everything under the sun.

After about 15 years of farming, my mother got really sick. She was living about 40 min away from us at Tenterfield at the time. Her husband was sick as well, so we were spending a lot of time down there. We had stuff all water on the farm at home and we didn't really have that much money on hand, so we just said, "Well, what if we give farming a miss from now on?" We're still on the farm but we don't actively farm it.

Before we had stopped farming, I'd gotten involved with the organic grower group in the Stanthorpe area. That took me to Stanthorpe Landcare, a group of landholders who were working together to manage environmental issues in the local community. I held a number of positions with this group: I became a representative of the organic grower group on the Landcare Executive Committee; I became Deputy Chair of the Committee; and then I became the Committee Chairperson.

When I was Chairperson of Stanthorpe Landcare, the Federal Government's Natural Heritage Trust 2 program, which funded regional natural resource management groups like ours, was coming to an end. I knew the changeover to a new program was going to be traumatic, because when they revise any of those things there's usually some people who miss out on funding under the new program, and some people who just don't like change.

At the same time, Kevin Rudd became Prime Minister in 2007, so we also had a new government. The Rudd government introduced the "Caring for Our Country" national funding program in March 2008. But by April 2008, it still hadn't started and they said it might not start until October 2008. At this stage, we'd had no Landcare Coordinator in the Stanthorpe area since about September or October the year before, for a variety of reasons; funding uncertainty was one of them, so we had no one to deliver the Landcare programs on the ground. I said to the Committee, "Look, it's April now, if they're going to wait until October, that's 6 months. We need a Landcare Coordinator. I know the area. What if I do the job for six months?" And they said, "Well, you'd still have to put in an Expression of Interest and have an interview". So

I stepped down from my role as Chairperson of the Committee and applied for the job as Coordinator on a temporary basis, until the funding sorted itself out.

I'm still the Stanthorpe Landcare Coordinator, but now the Queensland Murray Darling Committee (QMDC) funds my role. QMDC is a not-for-profit group that partners with some of the local Landcare groups in the Murray–Darling River catchment regions in Queensland. I assist in delivering environmental projects in the Stanthorpe area. It's my seventh year coming up tomorrow.

In this role, I work on a few projects. First, I work with blackberry, a declared weed under local law. QMDC has a blackberry incentive scheme that reimburses landholders who have to spend money to control blackberry on their properties, and I administer the program. The Local Council Officers for the Stanthorpe region, the landholders and QMDC all work together to try and get it under control.

I also work on the QMDC rabbit project. I really did fall into this role. Stanthorpe Landcare used to run blackberry control workshops. We would also have some rabbit work done at these workshops, just to make it a bit more interesting for the farmers. Blackberries and rabbits tend to go together. When QMDC first started funding my position, part of my role was supporting the QMDC rabbit project team through landholder engagement, by organising public meetings and helping landholders with their paperwork to claim a 30% rebate from QMDC as part of their rabbit incentive scheme. My role was fairly limited—I was waiting all the time for someone to contact me to set up a public meeting. I wasn't really given any message that I had to be more proactive. There were dedicated Project Officers running the rabbit project and I was just the person on the ground, the local person. After a while, a Project Manager job came up for the Stanthorpe area. I could see that we needed to start focusing a little bit more, and there wasn't really anyone else in the area or organisation that could do it. Then I just asked someone, "Look, you know… what if I was the Project Manager?" and they said, "That's a fantastic idea".

In my current role as Stanthorpe Landcare Coordinator, I also work with a range of community groups. I work with the Stanthorpe Rare Wildflower Consortium and assist in the coordination of a water-monitoring group. I engage with Southern Downs Regional Council, the local council in the Stanthorpe area, sitting on their pest management working group and working with the Local Laws Officers on pest management.

One thing I think I'm good at is running a good meeting, one in which everybody feels that they've had a good say and that their ideas, comments and suggestions have all been taken on board, and one that has a pathway to making a decision. You've got two elements there: you need a decision but you also need to have everyone involved in making that decision. You don't want people walking away from a meeting feeling unheard and unhappy. That is not a good outcome because these people become silent destructors; they will be sabotaging out there—unbeknownst to you—in whatever forum they find themselves, because they're wounded, and you don't let wounded people out the door. The way I run a meeting is I try not to let people get wounded. That's pretty well it.

To do that I use a range of techniques, some of which I learned through the way I was brought up. In my early childhood, my mother and I ran the house; we called my father one of the kids. There was a lot that had to be negotiated, so I think my mother brought me up to be a collaborator.

I've done courses, like the ones taught by Ian Plowman, a psychologist who works with government and community groups in Australia to teach them how to harness people's creativity. His meeting techniques are really quite useful, and I've taken a lot of his suggestions on board. For example, he has one handy hint for writing an agenda: if you write an agenda up and it's got points such as "stall at local market", when people see that point they think, "Oh, stall at local market" and then they start generating a whole lot of ideas, thoughts, opinions, attitudes and suggestions before anyone has had the chance to speak. His hint is that you can stop or minimise that behaviour by rephrasing the agenda item as a question. So the question might be, "Will we run the stall at the local market this year?" That's the question; it provides a focus, so straight away when they see it they don't fly off into all of these other ideas. Then when you get to that item on the agenda, you frame it in the same terminology and it clicks for most people.

Another good Plowman technique is that we don't talk about an issue right away, but I get everyone to just write down two different ideas pertinent to the issue at hand. I do a range of other things, but the "writing it down" technique works really well, and when something works really well in a meeting, I'm excited. When something doesn't work or when things are looking a bit poorly, that's when I really start racking my brain.

I also pay attention to people's body language. So while someone's talking about a particular thing, and we're at the discussion stage—not sharing ideas yet—I'll make sure that I'm looking around the room and seeing how people are reacting to it. If I see Jason sitting down there just constantly fiddling with keys, I know he's unhappy about something. So then I say, "Okay Jason, what thoughts do you have about this market stall, about whether we should have one?" Then he'll go off onto whatever has been distracting him for so long. It might be the fact that two and a half years ago they kicked him out of the market because he was drunk, or some bloody thing, and he's hated the people ever since. It doesn't really matter, but you really have to keep track of all these little weird things that concern people and affect their attitudes and behaviours.

Finally, I have a personal rule that no meeting goes for more than two hours, generally. After two hours, people tend to shut down. They get tired, and if you're going to meet any longer than that, you have to have a break in the middle.

My engagement around the issue of blackberries usually involves two kinds of interaction. One is workshops or field days, and the other is one-on-one or small group interaction, such as inspections. The thing about workshops is that people see the potential in a workshop, don't they? And you've got several people who contribute their own ideas, so you get a conversation that happens among the participants. That's one of the strengths of workshops, apart from the fact that with a workshop you can actually have a demonstration. You can have someone, for example fumigate a rabbit warren and show participants how to use traps. People get a lot out of that.

With the one-on-one's, you actually get the opportunity to talk to someone about their particular situation and some of the particular problems that they face, which they probably wouldn't talk about in a workshop.

Overall, what I find challenging is time management, my own personal time management; staying on track is sometimes tough. Every year, I work with the Committee to hatch a plan for keeping me on track. Then we go our merry ways and we're just really busy. I am getting better at it, but unexpected things crop up a lot. On the other hand, if I was really rigid and focussed, I probably wouldn't be very good at some of the things I do. So I try and see the flexibility as just having some unfortunate collateral damage in terms of meeting all of my outcomes.

What motivates me is protecting the landscape from adverse impact. I find stories about native vegetation and other landscape matters really interesting. When I'm talking to people in this space, various landholders or stakeholders, I approach them with a generous spirit. Some meetings can be very, very time-consuming, but basically I go to people with an open attitude. And whatever they say, I treat it as a serious question.

Chapter 16
What Can We Learn
from the Practitioner Profiles?

Abstract In this chapter we highlight the positive lessons, innovations, break-throughs, and insights gleaned from the practice stories. We make suggestions about how these lessons might strengthen the practice of community engagement for community-led pest management.

These lessons include:

- the importance of listening to build rapport and a nuanced understanding of the issue facing the affected community;
- having a clear drive to make a difference to social and environmental outcomes;
- the value of building common ground through sharing experiences with affected communities;
- the personal and professional benefits that come from a willingness to critically reflect and learn from experience (both positive and negative);
- the wisdom to strive for a sustainable balance between personal effort and community ownership;
- the political dimensions of community engagement and how these are reflected in practitioner choices;
- the importance of personal drive, communication, learning, and recognition of different perspectives.

The following sections explore these overarching lessons in more detail. In order to hear from the practitioners themselves, the analysis relies heavily on extracts from the profiles. However, the practitioners were not involved in the analytic process and we acknowledge that our view presents only one of many possible interpretations. This analysis is open for consideration, debate and correction by the reader. Adapting the analytic stance outlined by Peters et al. (2010), this chapter aims to "present what we have learned as lessons that challenge and enlarge our understanding of the nature, meaning, significance and value" of community engagement for invasive species management in Australia and hopefully "contribute a new thread in the conversation" about the way that practitioners go about their work with community (Peters et al. 2010, p. 316).

T. M. Howard et al., *Community Pest Management in Practice*,
https://doi.org/10.1007/978-981-13-2742-1_16

These lessons are the result of an appreciative, rather than a critical, enquiry. By appreciative analysis, we mean that the research team focused on uncovering the inspirational and positive directions that might support and strengthen engagement practice. This reflects a strategic choice to nurture an emerging awareness of the social and political dimensions of community engagement as an integral part of pest management.

This selection of themes presented is not exhaustive by any means—many other lessons and insights can be drawn—and we encourage readers to draw their own conclusions. This positive approach to the material respects the personal contribution of the practitioners, who have honestly shared their experiences without the protection of anonymity.

16.1 The Meta-Lesson: The Struggle for Change

We begin this journey with an overarching observation about the essence of the "work" that these practitioners undertake in their interactions with community members. Across the profiles, practitioners describe social change as a necessary strategy for achieving improved outcomes in the operational sphere of pest management. In Australia, the long-accepted norm that pest species are the responsibility of individual landholders is changing, as both government and non-government stakeholders begin to redefine the issue as a public problem requiring collective community action. The changing landscape of "shared responsibility" raises questions for practitioners who are often in mediating roles, working at the interface between government, industry and affected community stakeholders.

The desire to stimulate social change emerged through practitioner's experience of working with a range of stakeholders to address these issues of mutual importance. Most practitioners in this collection did not start their careers with the goal of achieving social change or, in some cases, working with people at all. Many of them came to engagement through seeing biophysical or technical solutions for an environmental or agricultural issue fall short through a lack of community action. The practitioners express a growing awareness of the influence of the wider social and political system (both current and historical) and an understanding of the role that all parties play in facilitating change. The practitioners and communities engage in the work of social change in two distinct forms: (1) attempting to innovate and try new forms of community engagement that might challenge familiar processes and procedures; and (2) working to change existing power dynamics, giving life to the ambition of "community-led" action for pest management.

Social change is not a simple objective, so we do not focus on specific achievements or outcomes; instead, we are curious to understand the different way that the practitioners themselves describe change and understand it within the context of their work with communities. Better understanding the change (how it happened, who was involved, and what helped or hindered people from coming together) offers a much richer picture of community engagement in pest management than species numbers can depict. Prompted by the bold observation of Mike Reid[1] (see Chap. 14) that the Victorian Rabbit Action Network (VRAN) was "redefining local and state politics", we began to listen for similar accounts of significant social change being supported and facilitated by practitioners.

As a public sector bureaucrat, Mike is aware of the challenge issued by participant-led design to established relationships between government and community stakeholders. For Mike, the struggle for change must be accompanied by a willingness to accept shifting power relationships. By establishing enabling settings for participant-led decision-making and knowledge sharing through the project's 'Learning Network' model, Mike is trying to subvert the usual balance of power by transferring action and agency into the hands of participants.

> People ask, "Well what does the Learning Network mean for managing rabbits? Is it just talk?" From what I can see, the Learning Network is a transformative process. Some people go back to their organisations and change the way that they work with community to establish rabbit action groups. Others go back to their communities, access funding, take leadership roles and organise events.
>
> …
>
> The VRAN and Learning Network have opened up the space for participants to do the talking. That's the best promotion the program can have. From my perspective, it's redefining local and state politics by reinvesting in people.

For Mike, as for other practitioners in this collection, this potential for community empowerment generates excitement and passion in his work with the community. The change underway is the result of participant efforts to work together in a supportive learning environment, however, innovation is often accompanied by instances of failure, as new ideas and ways of doing things are trialled. The challenge is to create safe policy spaces to reflect on the lessons that can be learned from failure and use these to improve the innovation.

The practitioners also faced the recurring theme: struggle in changing established patterns of both government and landholder behaviour, and associated power relationships. In this extract, Greg Mifsud[2] (see Chap. 13) reports that when faced with uncertainty, all stakeholders tend to revert to established ways of doing things. "It's not just the landholders who struggle with change," suggests Greg:

---

[1]Program Manager for Established Invasive Species in the Victorian Department of Economic Development, Jobs, Transport and Resources, Victorian Government.
[2]National Wild Dog Facilitator, IACRC.

> What really got to me at the beginning of this job was the government staff who would say, "That's not how we've done it in the past, so why would we do it like that now?" I couldn't really get any commitment from anybody to work with these [community] groups. ... Using a nil-tenure[3] approach to pest management planning is such a big shift for some landholders and, if someone isn't there to provide support and remind them why they're going down that path, they quickly revert to what they used to do.

For Greg, this is a sign that current policy frameworks fail to create the enabling settings required to support landholder action for wild dog control:

> The current ... trend of reducing regional staff and moving away from one-on-one landholder interaction has resulted in a large proportion of landholders not getting the information they need to understand feral animal control or that they have a requirement to participate in control programs under state legislation.

Within this struggle towards social change, the practitioners stories show the importance and the challenge of aligning ambitions for community-led action with the required enabling settings, such as adequate funding, time, resources and changes to established power relationships. Reading through the profiles, it is clear to us that community engagement is a dynamic activity that takes place in the context of constant change. This might be a change in the biophysical ecology; a change in the economic productivity of a particular enterprise; a change in the social dynamics; or a change in the practitioner's individual outlook. Each type of change has implications for the power relationship between affected parties. As you read these profiles, we encourage you to be attuned to these sometimes subtle, sometimes explicit, shifts in the balance of power and consider how these changes influence our understanding of community engagement as a practice.

Each practitioner describes different strategies and tactics to address power imbalances, such as meeting procedures or knowledge sharing forums. "I try to take all the individual and organisational agendas out of meetings, field days and workshops" says Darren Marshall, who goes on to explain that power can be wielded within groups of affected landholders:

> There is nearly always an individual in the community who is driving the agenda to control a certain species, and they will team up with the stakeholders that promote the same cause. This is dangerous, as it can often derail the focus of the larger community because they're just pushing their own barrow. It is important to address this, because without the entire community's ownership, coordinated control can't be sustained.

These are useful practice lessons in themselves. But sometimes these techniques may be inadequate for the challenges presented, as Brett Carlsson[4] (see Chap. 7) found out when working with family-run farms in central Queensland:

> Because his dad thought the weed wasn't a problem, no amount of convincing was going to change dad's mind .... That was the generational thing, saying, "We need to do it how dad used to do it while dad is still here".

---

[3] 'Nil-tenure' is defined in the Glossary.
[4] Wild Dog Coordinator for AgForce Queensland.

Brett's experience reminds us that in some cases, change may be slow. It may take generations and include factors beyond the immediate understanding or influence of practitioners, such as the interaction between social and psychological drivers. Practitioners, and their sponsors or employers, need patience to address these challenges and account for underlying community and individual values, historical experiences and tensions.

We conclude this section by clearly restating our understanding of the underlying motivation that animates the work of practitioners in this collection. It seems to us that each practitioner, through their efforts to effectively engage the community in collective pest management, is involved in the pursuit of social change. In line with our ambition to describe and articulate a professional practice of community engagement for pest management, we urge you to consider instances of change contained in the profile stories. The practitioners share their journey with us and through their reflections on both their personal and professional lives, help us understand their work in pursuit of change. We invite you to consider what the following practice lessons can tell us about the fundamental challenge for community-led action in pest management: the ability to affect change.

16.2 Practice Lessons

We pause here to introduce some overarching observations that emerged from our reading of the profiles. Each profile contained details of processes and favoured practices for community engagement. For example, individuals talked about how to run a meeting successfully, field a room of diverse interests, break the ice with a new community and ensure that the experience of community engagement would be a positive process. In this way, practitioners revealed a common commitment to the community stakeholders they engaged within their work, recognising their voluntary investment of time and effort through respectful procedures and good planning. This manifested as a willingness to be personally accountable for the way that community engagement might impact on participants, as Harley West[5] (see Chap. 15) described:

> When I'm talking to people in this space, various landholders or stakeholders, I approach them with a generous spirit. Some meetings can be very, very time consuming, but basically, I go to people with an open attitude. And whatever they say, I treat it as a serious question.

He went on to point out that this generosity of spirit also had a pragmatic dimension:

> You don't want people walking away from a meeting feeling unheard and unhappy. That is not a good outcome because these people become silent destructors; they will be sabotaging out there—unbeknownst to you—in whatever forum they find themselves, because they're wounded, and you don't let wounded people out the door.

---

[5]Stanthorpe Landcare Coordinator, Queensland Murray Darling Committee (QMDC).

Motivated by a common desire to make a positive difference in their professional life, several practitioners noticed that community members often felt unheard or isolated in their efforts to manage invasive species. Greg Mifsud suggests that:

> With the droughts and everything else landholders and farmers have to deal with, it can be a lonely place for them out there.

We observed that some practitioners also experienced this sense of isolation, either as a result of being the only person employed across a vast geographical region, or because of their changing understanding of their role within a bigger organisation.

Mentors and significant role models are referenced in each profile. Prompted by an interview question, practitioners nominate a range of influential characters: family members, research partners, colleagues and landholders. Sometimes these mentors provided intellectual inspiration by sharing books, theories and formal educational experiences; other times these were informal experiences, where mentors reminded practitioners to "keep it real" and ensure that their work remained focused on making a difference to on-ground outcomes. Some of these lessons might have come through informal avenues, like Jess Marsh's[6] (see Chap. 11) support network of others working in similar roles. Jess articulates the benefits of this network:

> … because even though we have similar roles, we approach things very differently. What I get from other facilitators is the understanding of what the position requires, new approaches and emotional support—which is really important!

Other mentoring relationships were deliberate and thoughtful, such as Lisa Adams'[7] (see Chap. 4) "apprenticeship" to two visionary leaders.

Thinking back on his career, Darren Marshall recalls: "I've always been lucky enough to be surrounded by experienced people and tried to learn from them". Darren's description highlights another important commonality between the profiles. Regardless of the form it might take, the practitioners all saw value in taking the time to observe, reflect and learn from those who had something to share with them. Some practitioners embody this commitment to lifelong learning by stepping into the role of mentor themselves. For example, Dave Berman[8] (see Chap. 6) sees his role as evolving to include:

> … doing science that is useful as well as having time to publish the important things that should be passed on to others. I hope to continue helping to protect Australia from vertebrate pests for at least a few more years and ensure that others benefit from my experience.

Passing on knowledge is another way that these practitioners demonstrate an evolving practice of community engagement. A career's worth of hard-won experience is a valuable asset for other practitioners, and the need for intergenerational learning is part of the rationale for this practitioner profile collection.

Before we leave this section, we want to make a strong statement about how practitioners might learn from these insights. The purpose of the collection is not

---

[6]National Natural Resources Management (NRM) Facilitator, IACRC.

[7]Biosecurity consultant; previous National Rabbit Facilitator, IACRC.

[8]Research Fellow, University of Southern Queensland.

only to articulate a community engagement practice for pest management, but also to strengthen this practice through stimulating critical reflection. Through this work of narrative enquiry, we draw attention to the way complex issues such as community action for pest management are identified, described, researched and addressed. In the profiles, we see many instances where practitioners are involved in this critique of issue "naming and framing". They describe strategies for accessing diverse opinions, learning through questioning and listening to the different ways that a "problem" might be understood and represented by various interests. Seizing opportunities to learn and reflect on what they experience in their community engagement efforts, the practitioners reinvigorate their personal practice through this "reframing", actively incorporating new perspectives on the problems or obstacles they encounter.

As a group, the practitioners share a conviction that different types of knowledge are essential to accurately naming and framing complex issues of community pest management. Rather than unquestioningly accepting the most visible or obvious definitions of these issues, we believe that community engagement practitioners will benefit from critically assessing the social, political and economic agendas involved in "naming and framing" public issues.

Inspired by the values of care, respect and learning evidenced in these practitioner profiles, we turn now to the practice lessons. Drawing on practitioners own words, these practice lessons aim to encourage reflection on the practical skills of community engagement, the underlying philosophy or values that shape this practice and the structural elements that support or hinder social change. You will notice some overlap and interaction between these lessons; we have not attempted to make artificial distinctions because we believe this interaction reflects the lived experience of community engagement for pest management.

16.2.1 Developing an Engagement Practice—Reframing "Community"

"Talking to people has never been a problem for me. I don't flick a switch to turn that on, it's something that I love to do", says Darren Marshall in his story of practice change. Despite his natural abilities and years of experience in the field, Darren had been "about ready to throw the towel in on working with the community" when he undertook a short community engagement course at Penn State University. Darren had been enthusiastically trying to engage landholders to participate in well-planned coordinated pig control action in central Queensland. He had become increasingly frustrated by the resistance he encountered and he had decided to "focus on ecological research—work with the animals and leave the people behind".

His experience at Penn State helped him reframe his work. He explains how this experience influenced him:

> The course was structured in such a way to show how community engagement is actually practiced. Previously, I was trying to bring information, relevant research and effective

planning tools *to* the community to assist in addressing the feral animal issues. The concept of developing research *with* community involvement from the beginning is not often practised by research organisations in my experience in Australia, and this changed how I thought I should work. This is where I started to realise that co-created projects that lead to a joint journey, (my own and the community's) with a mix of local experiences and knowledge, as well as research, could be extremely beneficial.

The theories, concepts and practical examples that Darren encountered allowed him to think about community resistance in a different way. He began to think about how local knowledge might be integrated with more familiar scientific approaches. Darren returned with a new commitment to his work:

I believe that when ownership and commitment of the community is strong, more effective strategies and management plans to reduce the number of feral pigs can be developed and their agricultural and environmental impacts reduced. In my experience, when community residents work together to address an issue important to them, they are then poised to handle other issues that may confront them going forward.

We pause here to emphasise Darren's last point which is particularly important to the aims of this collection overall. What he is suggesting is that the practice of community engagement may deliver benefits to the community beyond action on pest species. This way of reimagining the impact and outcome of engagement efforts taps into theories of community development that are seldom articulated in the field of pest management. Positioning Darren's work in the context of social change permits us to see his engagement efforts as part of a broader project of community building to create enabling settings for community-led collective action. This expansion of community engagement practice can transform the individual, community, organisational and institutional understanding of what community is and why engagement is important. Darren's experience at Penn State suggests that learning and drawing on the wide range of theories and methods developed in fields such as community development and social change can help structure and guide new practices.

16.2.2 Developing an Engagement Practice—Look for Opportunities to Learn

As we read Greg Mifsud's profile, a particular comment stood out. He was describing an informal interaction with a previously hostile farmer. Although Greg had worked hard to establish his credentials as a scientist and pest animal expert in the local community, there was still resistance from this particular farmer. Greg knew that without this individual's support, his effort to bring the community together for a coordinated wild dog control program was unlikely to succeed. Driving past a paddock one day, he saw this farmer and pulled over, hoping to start a conversation. This unplanned interaction turned into a productive conversation, as Greg describes in this extract:

I reckon we sat there for nearly an hour-and-a-half, on a log in a paddock, talking about the history of the region and how things had gotten so bad locally. We didn't end up talking much about dogs really—it was just an opportunity to learn about the culture of the community.

Greg illustrates another important element of community engagement practice that appeared in many of the profiles: the willingness to learn and engage with different sources of knowledge. This manifested in different ways but always reflected an underlying awareness that there is always new knowledge to gain. In Greg's case, this stemmed from a recognition of the unique context that each community faces in dealing with their pest problems and how influential this can be to achieving a change in behaviour. "In order to bring the community along for the ride" Greg says:

... I need to take the time to understand the community and the dynamics at play. It's hard to try and generate cultural and behavioural change when you don't live in that community or have a pre-existing relationship with the stakeholders.

This practice lesson supports recognition of listening and understanding issues from diverse perspectives as key elements of effective engagement practice. Practitioners suggest that asking questions and actively listening to the answers can help a practitioner learn about both the issue and the community context. For example, a practitioner entering a new community might ask: "Who is involved? How do they understand the issue? What has been attempted previously?" These questions help the practitioner understand how the issue is named and framed by an affected community. Seizing opportunities to learn helps practitioners build a better understanding of the dynamics that might be influential in their community engagement efforts.

Although some practitioners consciously ask questions and listen deeply to the answers, others also look for opportunities to learn from their own experience of community engagement. In this extract, Peter Fleming[9] (see Chap. 9) describes how he continues to gain new knowledge beyond his field of scientific expertise:

Reflecting on my experiences has been the way I've learnt about community engagement. I don't have any formal engagement training, and it's definitely been a limitation. I've just gone along and taken it in by osmosis; worked out what doesn't work from reactions and asking, "What did I do wrong there?"

Peter is describing a reflective practice of continual learning. Although he has an esteemed career in invasive species ecology, Peter knows there is always something more he could learn about the way he engages with the community. Peter turns his research skills inward to mine his experiences, both positive and negative, for insights about how to continually improve his practice, as he demonstrates in the following extract, describing his early experiences with a project focusing on wild dog management:

We had regular meetings with the steering committee and also some public meetings to tell people what we were doing. We would turn up at meetings and tell our story—it was pretty bad actually, the way we did it. A lot of the time, it was just messy. I was abused in a meeting; I was blamed for the telephone company leaving town, the banks leaving town, all sorts of

---

[9]Principal Research Scientist at the Vertebrate Pest Unit, NSW Department of Primary Industries.

issues. One lady was so distressed that she dumped everything on me as the government representative, and began to cry in the public meeting. I was dumbfounded but I've learned a lot about handling such situations since then.

Learning from disappointment or "failure" is an important part of growth and innovation. However, it also takes courage because there are few safe spaces for acknowledging these missteps. As Mike Reid noted earlier, this fear of failure can limit the potential for breakthroughs in new ways of thinking about, and actively engaging, community. Peter demonstrates that from his viewpoint, there is much to gain from critically reflecting on less-than-successful experiences:

> I learned that the first meeting [between researchers and community members] is basically letting people get issues off their chest—the first meeting gives people the chance to say their piece so that they feel like they're being heard. A lot of community frustration comes from the feeling of going around and around in circles. People have a meeting, they all get up and say what's been wrong for 10 years, and there's no action—nothing happens, and you come back 12 months later and do the whole thing again. You get embedded frustration.

> I also learned what to say in those meetings and how to say it. As a researcher, my job was to come and give people information. The people at the meeting would all just sit back and respectfully say, "That's very interesting, that's great, but it probably doesn't apply in this location because you did your work somewhere else". I realised that you have to frame the research around the question "What's in it for the stakeholder?"—whoever they happen to be. ... Be balanced; frame it for them but don't be afraid to present things that you've found. Don't hide stuff from them. Be honest; that's what you've got to be.

Reflective learning is not always a comfortable activity but, as these extracts illustrate, neither is the work of community engagement. Questioning is an essential part of learning about both the context and unique circumstances of each community and their pest issue; we suggest that practitioners can also develop skills in self-questioning that will enable them to continue improving their practice.

16.2.3   Developing an Engagement Practice—Learning to Listen

We want to understand how practitioners develop a community engagement practice in their work. As we previously noted, many of the practitioners came to engagement by seeing biophysical or technical solutions for an environmental or agricultural issue fall short through a lack of community action. Those practitioners who had received scientific training described a growing recognition of the particular skills and techniques that they found useful in addressing the social dimensions of these scientific problems. These extended beyond the realm of formal scientific method into more intuitive approaches.

For Barry Davies[10] (see Chap. 8), his engagement practice emerged from a combination of personal aptitude and a desire to improve on-ground outcomes: "it's just

---

[10]Previous Victorian Wild Dog Coordinator, Victorian Government.

about being able to communicate ... It's as simple as that. ... For me, engagement is about communication, empathy and personality."

Jess Marsh studied Marine Biology and recalls that her "entire university degree ... didn't touch on the human dimensions of anything!" Jess began her community engagement training on the job, picking up ideas and adapting her practice through trial and error. Initially, this was "terrifying" because Jess felt ill-equipped for the role, but over time her confidence began to grow. Jess began to recognise different versions of community engagement, as she describes in this extract:

> The position used to be called NRM Liaison Officer. I remember the old Facilitator saying in the handover that he was a Liaison Officer, not an Engagement Officer. This was my first insight into what it actually meant to "engage"—what was the difference between liaison and engagement? I think he thought he was doing his job by purely contacting and informing people—doing that very basic level of engagement.

Jess began to develop her own engagement practice. She incorporated a balance between formal and informal strategies to ensure that she was well prepared to address the unique context of each community engagement situation:

> Every problem and solution has a different history. This history can be positive or negative. Before I go out into an area, I try to look into what has already been tried, tested or implemented, why it did or didn't work, and the individuals that may have created conflict or driven the project. For example, I might get two calls about rabbits and both people might have the same issue, but the approach I take with each person is going to be different because they are different people. That helps me understand who I'm working with and what we're working towards.

In this extract, Jess emphasises active listening as an important engagement practice that ensures the community members she works with feel (and are) heard and that she is well informed about the subtle dynamics of the situation. Jess demonstrates her commitment to understanding both the issue under investigation and the particular context of the people she hopes to work with. She also reminds us that these efforts are time-consuming but, in her view, essential to developing mutual trust, two themes that recur repeatedly in these practitioner stories:

> There is little acknowledgement of how much time it takes to do engagement well and how much depends on personal relationships and trust. The way I see it, if I have built trust with someone, I can call them up and we pick up where we left off. If I say, "let's take this action", they're not as cautious.

While active listening is a necessary part of working to build trust, it is also a helpful technique for improving the relevance and on-ground application of scientific research. For Research Scientist Matt Gentle[11] (See Chap. 10), listening to community members is a necessary step in the research process:

> The engagement processes my team use are different from those typically used to get people involved in vertebrate pest control. We're not out there trying to persuade people to take a particular action. We gather ideas from discussions with people, not just facts; we collect information that provides insights into how people think, about what people perceive and

---

[11] Researcher at the Queensland Department of Agriculture, Fisheries and Forestry.

believe the truth to be. For us, it's important to understand people's needs and what we can do to help. Research would be irrelevant otherwise.

Matt makes a concrete link between the relevance of his work as a scientist and the way that his research can assist an affected community. He affirms a practical commitment to understanding the problem as the community sees it, rather than relying purely on a scientific expert diagnosis. In this extract, Matt describes how his problem-solving approach to gathering information can increase the likelihood of developing successful applied outcomes:

I think gathering more first-hand information from the target community is the way to go. Although a lot of the project ideas we work on come directly from people or local councils, often we are only receiving second-hand information about the issue. It's so important to know what people think is the real underlying issue. I don't want to do research the wrong way, or target the wrong questions. ... For example, if all the councils say," We have a problem with feral pigs", then I need to ask, "Okay, what is the real problem?" and go from there. I really have to talk to people about what they think the issues are, then they can be part of the solution—then they are the solution!

By integrating active listening, Matt's practice creates the potential for shifting power and responsibility for problem "naming and framing" from the expert to the affected community. Through the seemingly simple yet complex act of asking questions and actively listening to those people who have a stake in the problem, Matt opens up to the possibility that he can learn from other perspectives. He is entering into a research collaboration that it is not entirely within his control and therefore, he is also shifting the established power dynamics of the expert/non-expert relationship. By grounding his work in the unique context of the affected community and its needs, Matt shows how a simple commitment to listen can transform the relationship between community members and applied scientific outcomes. Matt's willingness to listen, to open up a dialogue and co-create a shared understanding lays the groundwork for a democratic practice of community engagement, a theme which we explore in more detail later in this chapter.

16.2.4 Developing an Engagement Practice—The Drive to Make a Difference

For research scientists with expertise in invasive species ecology, the human dimension of the work can often be surprising, as Dave Berman found when he started his Ph.D. investigating the abundance and management of wild horses in Central Australia:

Before I started the research, I had not imagined that it would involve working with people so much. I thought I was going to do a Ph.D. and simply learn about horses in Central Australia.

Wild horses are a well-loved species and acting to control these animals is often controversial. As a result of his experience in Central Australia, Dave began to see a link between undertaking good science and using this to communicate with people about the complex challenges of managing this charismatic species:

> When representatives of those welfare groups came out to Central Australia, we showed them the carcasses of horses that had died of thirst and starvation. In presentations I used hundreds of slides, instead of words or graphs, to communicate the science. I showed pictures of 80 carcasses around a waterhole in the Nineteen Mile Valley north of Kings Canyon, and said, "This is what happens if you do nothing". Showing those photos had a big impact.
>
> By engaging with the interest groups, we gained agreement on a common goal. The common goal was to reduce the number of horses. For the horse protection groups and animal welfare groups the reason was to prevent overabundant horses starving to death. For the cattlemen we were helping to increase cattle production and for the environmentalists we were protecting the native plants and animals and the soil. With this support we reduced brumby numbers by 70 to 80% across all of Central Australia during the time I was there.

This practical outcome reinforced Dave's drive to "make a difference" with his scientific research. This determination was in part inspired by the challenge from one particular landholder, as Dave recounted:

> He used to say, "What's the use of your Ph.D.?" That made me determined to apply the work I was doing. I wasn't just going to do a science project and publish it; I had a really strong drive to make a difference and as he put it, "be useful". I've had that same drive for every other project I've worked on since.

The drive to make a difference has also been a key guiding principle for research scientist Ben Allen[12] (See Chap. 5). As the son of a wildlife scientist, Ben had a long apprenticeship in the research trade, spending time out on farms learning the ropes of wildlife research. For Ben, getting out of the 'ivory tower' of academia is important for his applied research because it connects him with the interests and needs of the community:

> When I've got that applied science focus, I'm asking "How's this going to be good for the farmer?" or, "How can I save a threatened species without imposing on the farmer?" I think it makes me a better researcher.

In this extract, we can see how Ben deepens his research by approaching the problem from different perspectives. Like Dave, he adopts a stance of engagement with the interests and concerns of other parties. He tries to incorporate different sources of knowledge into his research, embracing the potential for improved outcomes because:

> The grassroots people, the people who actually manage the land, like farmers, they are so critical to doing a good research job. They often have a lot of knowledge for their particular part of the world, knowledge that a scientist is never going to capture in three or five years of being there.

The research scientists in this collection seek to engage their communities in both the naming and the framing of the "problem" they face. Motivated by their drive to make a difference, Matt, Dave and Ben recognise the value of incorporating

---

[12] Vice-Chancellor's Research Fellow at the University of Southern Queensland.

scientific and social values to understand the issue more holistically, not just for the additional knowledge this allows them to access, but because this willingness to engage around knowledge helps them communicate the aims and outcomes of their scientific research to the affected community. These scientists strive to bridge the gap between scientific expertise and community understanding because they see this as an essential pathway to achieving on-ground impact with their work.

Another dimension of the drive to make a difference can be found in Mike Reid's description of the personal motivation he draws from his work with community members. He starts by describing his involvement in implementing a model of community-led action for blackberry control in Victoria:

> One of the things the VBT [Victorian Blackberry Taskforce] identified early on is that—although we know how to control blackberry from a scientific perspective—if we started looking at blackberry as a people issue, it opened up a whole set of different solutions. And that was our mantra: "We're going to define this as a people issue: blackberry control is more than science".

When he was working out in the field with his community partners, Mike felt sustained and supported in pushing for this new perspective to an entrenched weed problem. However, this reframing of the blackberry problem as a social issue proved harder to implement in his government workplace. Co-workers felt that he was challenging the established way of doing things and this led to resistance. "It wasn't always smooth sailing", Mike tells us:

> There were tensions. In the early days with the VBT, I was so passionate about and didn't understand some of the internal departmental politics and power. I'm a very emotional person. Although we had support at a high level in government, a lot of the compliance staff would say, "No, you're not coming into this community". For me it was like a rugby match. I had the ball, and a line of compliance people to push through. I felt isolated in the [Ag] Department, pushing this approach, and that was hard.

His passion for making a difference encouraged him to persist and he began to see more systemic changes in his workplace. This has resulted in increased funds and support for community-led action and "witnessing those changes" has been really moving for Mike, who said he can now "go home at the end of the day and feel like I've made a real difference".

16.2.5   Developing an Engagement Practice—The Practical Side of the Business

Working in pest management often requires working in rural and regional areas. Rural communities are unique workplaces, particularly in Australia, where long distances between properties can make face-to-face interactions difficult. Often the "community" that practitioners are seeking to engage are farmers or other land managers who are used to being self-reliant and may be suspicious of "outsiders" such as community engagement facilitators, extension providers or, as Ben Allen pointed out, university researchers. This raises an important issue for those practitioners whose role requires them to interact with rural community members: How can they establish positive, respectful relationships that enable them to meet their community engagement objectives? As we read through the profiles we saw that a key strategy for bridging this "insider-outsider" divide was being able to demonstrate that the practitioner understood the practical side of living in a rural community. Many of the practitioners drew on their early experiences growing up in farming communities or rural areas to establish credibility with their rural stakeholders. Brett Carlsson describes how he relies on his practical knowledge of farming in central and western Queensland to do his engagement work:

> … if I didn't have the practical knowledge I've got about pest management—if I just had the science knowledge—I don't think I would be as successful as I am in my role because landholders soon find out whether you know what you're talking about. … I don't know how many times I've pulled up to go and talk to someone about pest management, and first I've had to help them unload cattle off the truck or finish branding in the yards. Once I show them I can do that practical side of their business, then we sit down and I show them the pest management business.

For Brett, establishing his credentials as an "insider" with a long-term commitment to the area helps him gain trust. Like other practitioners in this collection, Brett worked hard to become part of the social fabric in these communities and saw this as an essential ingredient in his engagement practice:

> … I know a lot of the landholders socially from living in Longreach for 10 years before I moved to Mareeba. If I had never lived in Longreach and I was just a Mareeba guy going down to help them put their control programs together and coordinate aerial baiting, I would not be well received at all. But, because I lived in the region for 10 years, I know their plight, I understand their position. To me, it is critical to how my role works; if I hadn't lived in Longreach, this project wouldn't have been successful. I guess that's why they gave me the job.

Echoing Jess Marsh and Greg Mifsud, Brett reminds us that establishing these connections can be time-consuming because it requires human interaction:

> It's nothing for my phone to be ringing at 8:30 pm for a chat about dogs. I answer it because that's my job. I know they appreciate that and they tell me that they appreciate it, so it's worth the effort.

Many engagement practitioners in this collection have rural backgrounds. They continue to draw on their early experiences of country life to demonstrate that they know about the "practical side of the business". For example, Barry Davies explains how growing up in a farming community in the wheatbelt of Western Australia equipped him with the practical skills he continues to rely on:

> If I happen to be driving past [a landholder's] place and he's fencing on the road, I will just stop and have a chat. If he's got some stock in the yard and he's drafting sheep, I will jump in the pens and give him a hand. What better way can you be seen as being a normal person, a human being?

This quest to establish common ground can lead practitioners to take risks, as Dave Berman found when he moved to Central Australia to undertake research into wild horses. In this extract he explains the lengths he went to gain a key landholder's trust:

> … [the landholder] gave me a hard time for at least 18 months until one day I turned up to conduct horseback transects through the rugged part of the property and I came across the Turners mustering cattle into a yard. I asked if I could help them. They said I could but they were nearly finished. So I jumped on one of my horses, bare-back (no saddle) and rode out to help. Although they were close to the yard and almost had the cattle in the gate, the cattle had other ideas. The cattle split up instead of going into the yard and went in all directions. I galloped after one of the steers, jumping gullies and crashing through the scrub. I managed to shoulder the steer to a standstill and helped bring him back to the yard. At last I had demonstrated that I was useful and that I appreciated the difficulty of their job. We gained mutual respect that day and I was trusted at last. In fact from then on I was treated as one of the family.

A common thread through all of these accounts is that successful community engagement is characterised by action—in these extracts we see practitioners taking steps to actively demonstrate their relevant experience, commitment and capacity to tackle issues in their community context, rather than approach them from a distance. They show us that community engagement relies on human interaction and a willingness to try and bridge the distance between the practitioner and the affected community. Practitioners regularly emphasise their active efforts to build mutual trust with their community stakeholders as a crucial element of their practice.

What lessons can practitioners without rural backgrounds draw from these accounts about understanding the practical side of the business? To answer this, we turn once more to Barry Davies. Although Barry identifies as an ex-farmer, he doesn't see this as the only pathway for establishing credibility and trust with rural communities. In this extract he suggests that authenticity and integrity go a long way to establishing trusting relationships:

> I've learned that if you're not honest and upfront with the community about what the rules are, what the government can and can't work on, you might get away with it for a while, but not forever. It's important not to make promises that can't be kept, because if you don't deliver on it, then you lose all your credibility. That's the way to lose the community's trust, and rightly so. I believe that if you show integrity, that you're there for the right reasons, you will do well.

Barry reminds us that if we understand community engagement as a relational activity, one that rests on regular interaction and interpersonal communication, a practitioner can demonstrate integrity and authenticity through their approach to the work.

16.2.6   Developing an Engagement Practice—There's Only so Much You Can Do

How does a community engagement practitioner stay motivated when their best efforts are not enough to achieve the on-ground outcomes they hope to inspire? This question emerged spontaneously from the practitioner interviews. As we read and re-read the accounts, this struggle to stay motivated revealed some lessons that seemed important to share in our aim to describe and build a practice of community engagement. It would be easy to gloss over these messages of frustration. But the more we reflected on what these uncomfortable insights might mean, we began to understand that it is important to canvas both the ups and downs of community work. In the end, it was the practitioners who guided us. We took particular inspiration from Mike Reid, who aspires to what he calls a "warts and all" approach:

> In community engagement, it's important to create safe spaces to fail and be able to talk about that failure. Sometimes you might have to make hard decisions or compromise on one of your values in a meeting. I want to be really honest and transparent about that.

Greg Mifsud also articulates the challenge of keeping a balance between the aspirations and the realities of working in community engagement:

> These facilitation roles are difficult because it can take a lot of time and effort before any change occurs. In this job, you very rarely get closure, and it's often hard to see whether you are making a difference.

In this extract, Greg makes the point that community engagement can be a challenging profession. The personal drive to make a difference that inspires most of the practitioners in this collection can make it difficult to maintain strict work-life boundaries. As noted by several practitioners, there is a significant time commitment to building interpersonal relationships which can manifest in hours of travel time or late night phone calls. For example, when emotions about wild dog attacks or rabbit impacts run high, practitioners are caught between the needs of the community and the professional limits of their job description. If on-ground outcomes are not fully realised, the practitioner might begin to question whether this personal investment is really worth it.

For the practitioner to maintain perspective, it may be necessary to reframe their community engagement objectives. Drawing on his years of experience working with landholders across the country to manage wild dog populations, Greg offers an alternative way of understanding his role, suggesting:

> … just because a community engagement program doesn't deliver the expected result or outcome doesn't necessarily mean it is a failure. When I first started, I thought that if the group didn't get organised that people would frown upon me because I hadn't achieved my purpose. I set numerical targets each year for the number of groups I was going to work with or write plans for. I don't do that anymore, because if one group doesn't come on board then I don't reach my milestone, and it is no slight on my capacity if a group or person chooses not to participate.

In this extract, Greg demonstrates how he repositioned his thinking about his role and its objectives. Greg recognises that there are limits to what an individual effort can achieve. He has positioned his role within the broader context of a community effort. As a result, he has found a way to sustain his practice by acknowledging that the community must also take action if the real value of his role is to be realised.

16.2.7 Developing an Engagement Practice—Combining Theory and Practice

As we read through the profiles, we began to see how some practitioners draw on a combination of political theory and personal values to structure their work with community members. For Lisa Adams, seeing herself as working within a broader tradition of civil society is important. In this extract, Lisa describes how she integrated political principles into the design and implementation of a community-led rabbit management project:

> I asked "What influences your decisions and actions? Where do you get your information from? Who do you trust?" I asked for permission to use particular extracts and compiled them into a 20-page briefing note which showed what the farmers were thinking, local government, state government, environmental consultants and so on, in their own words. This is an example of how democratic practice has shaped this project.

Lisa goes on to explain how an emphasis on democratic principles caused some discomfort for those who expected a more familiar pathway of community engagement. In particular, Lisa found that:

> The word "democratic" gets easily lost. Throughout the project, when other staff or stakeholders were talking about how the project was designed and what it hoped to achieve, the word "democratic" would just drop off, it would go missing. I would say, "We've got to put that word back in there. That word is important". People would ask, "Why? What does it mean? How does it affect the project?" The way that I approached the project was driven by this idea of democratic practice.

We see her commitment to the terminology of democracy as a motivating force underpinning her personal practice of community engagement. By staking her project's claim to democracy, Lisa created an expectation amongst both her community participants and her governmental colleagues. The word "democracy" became a benchmark for the entire project effort.

If we read closely, we can identify some of the elements that make up Lisa's democratic principles and how these elements translated into her way of approaching the project design. For example, Lisa states her belief in the value of local knowledge, describing the process for developing an initial briefing note addressing rabbit management:

> Understanding the wisdom of people in the communities who are dealing with the problem is important. The briefing note was compiled from interview extracts so that different types of knowledge were given equal space for consideration. ... Everyone saw that it was an

important problem from wherever they sat within the system. There was no favouring one knowledge over another; it was collaborative, not competitive.

The democratic practice Lisa describes in this extract has a particular quality to it—a willingness to embrace different types of knowledge, not just to gain more information, but to share power amongst all of the key stakeholders. This approach moves away from an expert-dominated way of naming and framing the issue, to create a platform for more equitable participation. Lisa continues to describe the outcome of this process:

> The briefing note had given everyone access to the same base level of knowledge and information. There was a shared understanding, which meant they could move straight to deliberation. They were able to move straight into thinking about rabbits as a complex system. They were able to move beyond their individual interests to think about the collective problem.

Lisa's commitment to sharing knowledge, and therefore power, amongst the rabbit community, had a profound effect. It created a common understanding of a complex issue without silencing or avoiding different perspectives. Through a determination to democratise the process of project design, Lisa illustrated a possible pathway for achieving collective action. There are other useful practice lessons in this example but we would like to highlight one that is of direct relevance to our aims with this collection. We see Lisa's integration of a particular theory of democratic practice as a significant progression in the articulation of what community engagement is and, even more significantly, why would we try and do it differently.

First, let's consider the *what*. Community engagement is often used to describe a range of activities or efforts that have the "community" as an intended audience. It might include marketing campaigns or specific events; sometimes it is an invitation to participate in a political process. Most commonly, community engagement is considered a tool to attain community acceptance or support for a particular, predetermined outcome. Lisa's story of the rabbit project shows us another possible vision of what community engagement might achieve: the creation of common understanding and increased potential for shared ownership of complex issues. These potential outcomes require hard work and an explicit commitment to try and share power, reduce centralised control and embrace unexpected results. As our practitioners have suggested, this empowering vision of what community engagement might achieve can be challenging, particularly when it suggests a change to established organisational and bureaucratic ways of doing things.

A crucial question emerges from this practice lesson: How can practitioners hold onto their determination to do things differently? Lisa's profile suggests that developing a personal philosophy of community engagement is one of the most crucial steps in learning to do things differently. Identifying and articulating the values that underpin an individual' s engagement practice helps to strengthen the *intention* of the work. For some practitioners, this may take the form of engaging with particular theories of community development and political economy; for others, it may mean reaffirming ethical behaviours and values that have been influential throughout life. Regardless of where these principles and values come from, we suggest that it is the process of consciously articulating them that adds discipline and rigour to the *practice* of community engagement.

This brings us to the why: *why* would any practitioner try to implement a different way of thinking about and practising community engagement? We suggest that the desires and objectives that practitioners raise in this collection provide the answer to this question. For example, Barry Davies' desire to help farmers "gain more control over what they [do] on their farm or station"; Greg Mifsud's endorsement of nil-tenure planning for wild dog management because it brings "everyone into the room" and gives them "equal say in the decision-making process"; and Darren Marshall's belief that "effective community engagement is about having landholders involved from the start". Other practitioners such as Matt Gentle, Ben Allen and Dave Berman, achieve a similar outcome in their approach to scientific research. Each of these practitioners demonstrates a practice committed to *sharing power* through collaboration and inclusiveness. Their experience has convinced them that addressing complex public issues such as pest management requires an inclusive approach to naming, framing and acting in order to increase the likelihood of community action. They embody principles of democratic practice, without the explicit reference to the underpinning theory that we see illustrated in Lisa's story of rabbit management.

We suggest that incorporating democratic principles in the design and implementation of community engagement may deliver benefits beyond action on pest species. Practitioners are encouraged to see their engagement efforts as part of a broader project of community building to create enabling settings for community-led collective action on a range of challenging and complex issues.

16.3   Learning from Practice Stories

This returns us to the main aim of this collection of practitioner profiles: to describe and articulate a professional practice of community engagement for pest management. These practice stories present an example of how we can learn from listening to each other and suggest that, if we take the time to tell our own story, we may be able to draw lessons from these experiences too.

Reading the individual profiles illustrates the range of different approaches that each practitioner takes, informed by their personal and professional experiences over the course of their career. Although we have spent time examining this collection for useful practice lessons, it would be foolish to suggest that these are the only lessons that can be identified and equally as foolish to set them out as recommendations that will work for all practitioners in all settings. We prefer to reinforce the intention and philosophy that is captured in Matt Gentle's reflective question, "How can I do better next time?"

We encourage readers to approach these stories of community engagement practice as an opportunity to learn about themselves and their practice. By adopting a learning stance, it is possible to draw practical lessons from these stories of success and disappointment. This increases the capacity for individual practitioner resilience and helps sustain your community engagement efforts going forward. It also helps develop knowledge to share with others, increasing the capacity of practitioners to learn and mentor each other to develop a professional practice of community engagement for pest management. In closing this section, we return briefly to Lisa Adams, who reminds us:

> If we are humble in what we know and what we don't know, we can reach out and learn from others. We cannot carry the whole on our own—we have to work together.

Reference

Peters, S., Alter, T. R., & Schwartzbach, N. (2010). *Democracy and higher education: traditions and stories of civic engagement. East Lansing*. MI: Michigan State University Press.

Part II
Wild Dog Groups—Three Case Studies

Chapter 17
Introduction: Wild Dog Management Groups

Abstract In this section, the social dynamics that underpin community formation in wild dog management groups are revealed in three case studies of wild dog management groups. The narrative approach allows us to see that:

- Groups create shared stories about the problems they face and the solutions they pursue; and
- These stories include details about the process of group formation and action for wild dog management.

In these stories, the affected community is shown to be:

- resilient to change,
- firmly located in the landscape,
- persistent over time, and
- operating in a highly context-driven climate of social relations, power dynamics and historical tensions.

The stories help us understand:

- the emotional dimensions of wild dog management;
- the constraints on community capacity to act;
- the pivotal role played by community leaders in generating and sustaining community efforts;
- the expression of power through the creation and sharing of knowledge; and
- how issues of community pest management are framed, including concepts of both success and failure.

In this section of the book, we turn our narrative enquiry from the individual to the collective, with an examination of community wild dog management groups in Australia. Through three case study narratives and accompanying analysis, we explore examples of community action to collectively manage wild dogs in a range of different economic, political, geographic and social contexts. These case studies

© Springer Nature Singapore Pte Ltd. 2019
T. M. Howard et al., *Community Pest Management in Practice*,
https://doi.org/10.1007/978-981-13-2742-1_17

combine multiple perspectives, drawn from individual experiences of wild dog management groups, to build cohesive stories of group formation and operation. Case studies were located around Australia to capture the diverse context within which wild dog management occurs. Locations included Mount Mee in southern Queensland, Ensay-Swifts Creek in the alpine area of Victoria, and the Northern Mallee region of southern Western Australia (see Fig. 17.1).

The case studies investigate how specific communities responded to the threat of wild dogs. Case study selection was guided by the research interests of the Invasive Animals CRC, the National Wild Dog Facilitator and a range of state and local partners. Through a process of dialogue, our research partners identified a need to understand more about the interaction between wild dog impacts, funding and coordination efforts, and community response. Motivated by the national policy drive to implement shared responsibility models for the management for widely established pest species, in this case by stimulating and supporting community-led action for wild dog management, the case studies aimed to increase knowledge about the social dynamics that underpin community formation in wild dog management groups. Groups of landholders have been identified as key to achieving sustained invasive species management in rural landscapes (Marshall et al. 2016). These groups are the interface between individual landholder experience of pests like wild dogs and top-down government responses through legislation and policy implementation. Most

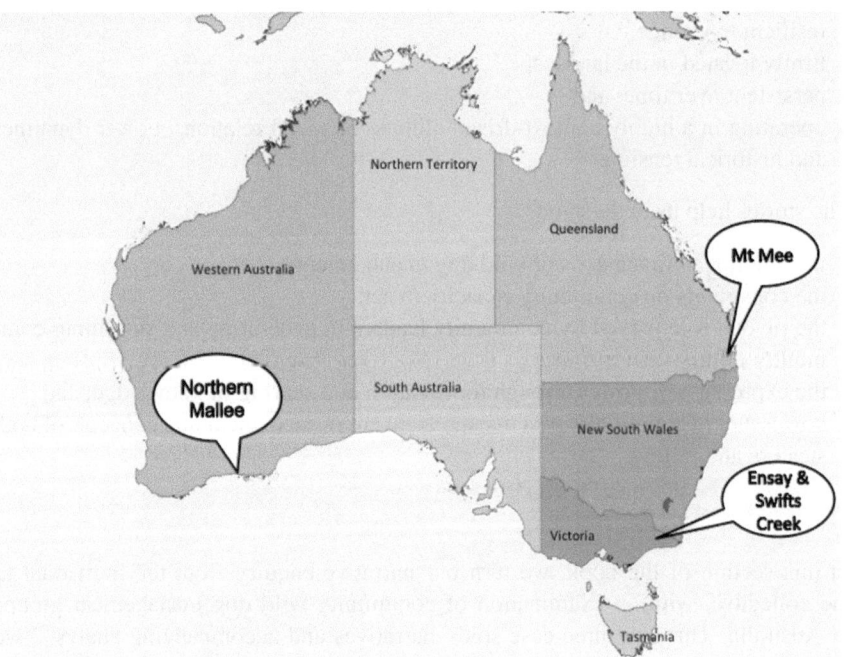

Fig. 17.1 The map illustrates the location of the three case study sites

importantly, these groups hold useful contextual knowledge that affects a community response to a particular invasive species issue (Everts 2015).

Landholder groups are suitable for narrative enquiry because they create forums for shared experience and social interaction. Group members regularly engage in the telling and re-telling of personal stories about the problems they face and the solutions they pursue. Through sharing and reaffirming each other's stories, groups create a unifying narrative that makes sense of group formation and action while also illuminating broader social, cultural or political dynamics (Fine 2012). By unpacking the details of particular case study groups, including context-specific circumstances of each, useful insights about community formation and collective action for wild dog management are revealed. Through accessing a range of individual stories and considering them in relationship to each other, we developed both a collective story and an accompanying analysis that engaged directly with the shared experience of the group.

Before delving into the narratives, we would like to share some of the analytical challenges we encountered as we strove to make meaning from the collection of first-person perspectives on the emotionally and politically charged issue of wild dog management. The broader principles of narrative enquiry and our specific intention in applying this to our work have been outlined in Chap. 2 of this book. The specific methods used to gather data and develop the case study narratives have also been described in Chap. 2, however, there are some additional details that require elaboration here.

As we have emphasised throughout this book, the narrative enquiry does not strive for scientific notions of reliability and generalisability. Based on individual recollections of personal experience, a narrative will always present a particular and partial account that may not resonate with all readers. Narrative enquiry is based on the creative acts of storytelling, listening, re-storying and meaning making that underpin the social construction of knowledge (Clough 2002). In these cases, we collected interviews from community members who had actively participated in the specific wild dog response or those who were directly impacted by these activities. As a result, the sample was skewed towards community members and individual landholders rather than policy or program designers. This reflected our research interest in understanding how those people who formed the affected community described and made sense of their experience, rather than an attempt to develop a precise historical or procedural account of the case. As a result, these case study narratives present the reader with a community-centric viewpoint of community action, embedded in the broader social and political context described by the interview subjects.

In recognition of this constraint, we invite readers to engage with our interpretation, raise challenges and consider alternative versions of the story. Indeed, this process has already begun. In the following paragraphs, we share some examples of how our interpretation has prompted critiques and retellings. Through these examples, we hope to illustrate that subjectivity is a fundamental aspect of storytelling and remind the reader that each individual will have a different "take" on the events, depending on their experience and position in the case (Clandinin and Connelly 2000). This emphasis on experience helps us engage with the complexity and apparent unpredictability of human behaviour in relation to community formation and collective action for wild dog management. As you read these narratives and the accompanying analysis, you may notice tensions and contradictions that are allowed to coexist. This reflects a deliberate decision on our behalf, as researchers and analysts. In our construction of these narratives, we have worked to craft a readable story that embraces these inconsistencies, rather than preferencing a smooth and inherently logical account. Incorporating these different perspectives into the same story encourages us to listen to the contrasting accounts and understand that each reflects a particular reality, a lived experience, that is influential in the overall portrayal of community action. For example, there is a tension between stories of community-led action and reliance on government funding, coordination, or support that emerges in all narratives. This tension exists because individual interview subjects simultaneously expressed both versions of "reality" in their story of community action. Rather than try to smooth this away, we have retained the contradictions because to our way of thinking, they speak to the active and ongoing tussle between the ideals and realities of community-led action for wild dog management. From our perspective, these divergent viewpoints demonstrate how understanding and engaging with subjectivity is an important capability for those working to stimulate community-led action.

During analysis, we became sensitive to the challenge that our community-centric approach might pose to those research partners involved in policy and program design. To test our interpretation, we sent the emerging narratives to policymakers in each case study jurisdiction. While the Western Australian response was largely positive, finding "useful insights" in the case narratives, the Victorian response was more qualified, acknowledging that "like all stories, there are always alternative perspectives" that they felt were not adequately captured in the story. A national industry perspective queried why particular influential employees did not appear, suggesting that community members had overlooked how crucial these positions had been to stimulating community action in each case. These readers were alert to the likelihood that uncritical bias might creep in "due to the narrow (myopic) view of the actors involved in this story".

These responses provided helpful detail about the chronology of events, which were incorporated in the final narrative where appropriate. However, to avoid privileging this perspective, we did not change our analysis. Rather, we decided to include some detail about this interaction between the research and the reader here, to demonstrate how readers might take up our invitation to critically engage with our interpretation.

During this research dialogue, several inconsistencies and points of difference between the community and policy account were highlighted. For example, one reader acknowledged that:

> ... while some elements in the narrative are important (tensions within and between communities and government), there are some perspectives missing through being too close to the issue (there are always more sides to consider).

The analytical challenge posed by these responses reaffirmed our commitment to the narrative method by illustrating how each individual views a story from their specific vantage point. For example, those coming from a policy perspective highlighted the distinction between community activities and policy processes; they disputed landholders' accounts which collapsed these categories together and failed to adequately recognise when institutional or political support had been provided. For example:

> … [you] need to be clear that the community views government as a singular when it is actually much more complicated. Much of the heavy lifting was done via policy.

Sometimes the critique supported our interpretation of the underlying social dynamics revealed through the narrative analysis. For example, in the instance where one community group saw themselves as unfairly disadvantaged in comparison to their neighbours, the policy response pointed out many instances of equitable treatment that had not been recounted by the interview subjects, suggesting that the group "appears to [have] a bit of a 'chip on their shoulder'". This reaffirmed the conclusion of the narrative analysis which found that an emotional undercurrent of disadvantage had influenced the group's response to wild dog management.

Another response provided detail about the historical context of one case which had not been mentioned by community members during the interview process:

> It is important to understand the history of the program. The government, as a landholder, had an obligation to manage dogs. The consequence of the court case meant that the primary aim of the government operations was to meet the statutory obligation of "reasonable" control. The landholders have the same obligation, but blame government for the dogs perceived to be coming out of public land. The other counterfactual is the government's knee-jerk reaction to the court case. The government essentially shot itself in the foot by creating a welfare type dependency by community on government for wild dog management.

From a methodological point of view, this example points to the importance of building adequate field texts that support the development of rich and nuanced narrative accounts (Clandinin and Connelly 2000). Being aware of significant legal and policy circumstances helps to ground a narrative in the broader cultural context. This can increase our understanding of how community expectations have evolved over time within the highly politicised context of wild dog management in Australia. This was reinforced by another policy respondent, whose awareness of historical patterns alerted us to influential dynamics underpinning community-led action:

> In the past government would step into coordinate species control and that built an expectation about the government's role that is no longer the case.

Another critique raised questions about a perceived lack of acknowledgement of the external support provided to these wild dog management groups, in the form of government officers or industry facilitators. We suggest that this reflects the tendency of community-centric narratives to privilege the insider over those that seem external to the group. When these external roles are recognised in the narrative, their characterisation may not be as accurate as policymakers or industry bodies might hope. Although it is comforting for governments, industry and other funding bodies to read positive accounts of their efforts, we cannot shy away from the challenge that negative responses pose to our understanding of community-led action. It may be helpful to consider these challenges in the historical context that our policy respondents were so alert to. Community-led action requires a renegotiation of historical power dynamics and this may present as a challenge to authority or a lack of acknowledgement of external support (either policy or practical). A persistent community narrative of "us against them" serves to unite community members in both their struggle to manage wild dog threats and their struggle to realise a version of community-led action that is authentic, effective and sustainable over time. We urge our readers to recognise these tensions between authority, dependency and community action as part of the messy, gritty reality of the community-led response.

We believe that these case study narratives contribute a fresh perspective to questions of community-led action for wild dog management in Australia. This perspective captures first-person experience of participants in a collective action effort, privileging a community-centric telling of the story in an attempt to provide a useful corrective to the stories usually told through policy or program evaluations and technical reports. Through a process of narrative analysis, the community is shown to be resilient, firmly located in the landscape, persistent over time, and operating in a highly context-driven climate of social relations, power dynamics and historical tensions. Our intention in this introduction has been to acknowledge both the potential and the limitations of narrative enquiry in order to encourage you, the reader, to take an active stance of critical reflection in your approach to our interpretation. As one of our policy respondents pointed out "there are always more sides to consider"—we invite you to add these perspectives to your interpretation as you read.

The rest of this section is arranged as follows.

Each case study is comprised of three parts:

- the introduction, which provides a brief synopsis of the narrative analysis and some useful physical and institutional context;
- the case narrative, which combines first-person perspectives to tell a story of wild dog management; and
- the narrative analysis which draws out what we see as the interesting and influential dynamics of each case.

The three case studies are followed by an integrative meta-analysis, which draws out overarching insights that contribute to the broader conversation of community-led action for wild dog management, with application to both the Australian context and beyond. This meta-analysis concludes the wild dog narrative section and leads us to the final section of the book, where we present conclusions from our narrative approach to community pest management.

References

Clandinin, D. J., & Connelly, F. M. (2000). *Narrative enquiry: experience and story in qualitative research*. San Francisco, CA: Jossey-Bass.

Clough, P. (2002). *Narratives and fictions in educational research*. Buckingham, England: Open University Press.

Everts, J. (2015). Invasive Life, Communities of Practice, and Communities of Fate. *Geografiska Annaler: Series B, Human Geography, 97*(2), 195–208.

Fine, G. A. (2012). *Tiny publics: A theory of group action and culture*. New York, NY: Russell Sage Foundation.

Marshall, G. R., Coleman, M. J., Sindel, B. M., Reeve, I. J., & Berney, P. J. (2016). Collective action in invasive species control, and prospects for community-based governance: The case of serrated tussock (Nassella trichotoma) in New South Wales, Australia. *Land Use Policy, 56,* 100–111.

References

Chapter 18
Case Study: Mount Mee Wild Dog Program—Moreton Bay Shire, Queensland

Abstract The Mount Mee wild dog case study describes a local government program that supports local landholders to participate in coordinated control. The Mount Mee example shows how local government leadership can support landholders to increase participation in wild dog control. Making it easy for landholders to participate is identified as the most important ingredient for success in this case. The local government program demonstrates a commitment to respectful communication with landholders and develops procedures to build relationships across the region. Council staff use local wild dog data to engage landholders in conversation, integrating scientific expertise with locally produced knowledge. This approach has created a safe and embracing culture that is shown to nurture community participation. By reducing the administrative and regulatory burden on individual landholders and community leaders, local government enables responsible and motivated members of the community to focus on building relationships, sharing information and creating a norm of civic duty and participation amongst their neighbours. The local council's willingness to act for the public good creates political and social capital that leads to a generally supportive and encouraging atmosphere for wild dog control. The needs of the farming community, local government and state agencies have become aligned with the aspiration to make a civic contribution to the wider public good.

18.1 Case Study Context

18.1.1 Geographic and Physical Context

Mount Mee is a peri-urban community located in the Moreton Bay Regional Council (MBRC) area, between Brisbane and the Sunshine Coast in the hinterland of southeast Queensland. It is approximately 90 min driving time north-west of the Brisbane CBD. See Fig. 18.1.

Mount Mee is part of the D'Aguilar Range, and is bounded to the west by the D'Aguilar National Park. The Park includes the Mount Mee Forest Reserve, which contains a variety of rainforest and eucalypt forest, and is a popular tourist destination for bushwalking , camping and four-wheel driving. Other tourist attractions include

© Springer Nature Singapore Pte Ltd. 2019
T. M. Howard et al., *Community Pest Management in Practice*,
https://doi.org/10.1007/978-981-13-2742-1_18

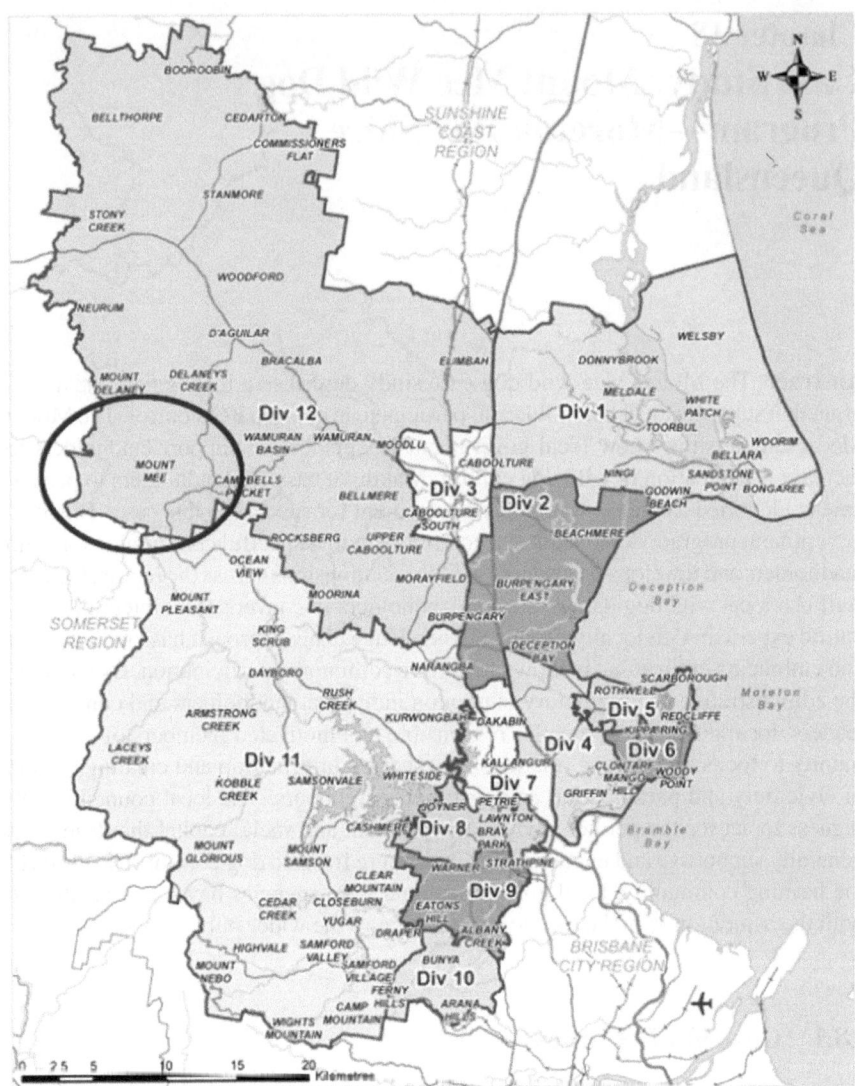

Fig. 18.1 Mt Mee (circled) within the Moreton Bay Regional Council. © Moreton Bay Regional Council 2017 (Reproduced from Moreton Bay Regional Council 2017)

the remains of an early sawmill, guided tours to learn about the history of forestry and logging in the area, and local restaurants and bed & breakfasts (Tourism and Events Queensland n.d.).

Early primary production in the Mount Mee locality focused on forestry, with timber getters being attracted to the area from the late nineteenth century to source the red cedar and other in-demand timbers then abundant at Mount Mee. Dairies and banana plantations followed as key forms of agricultural production (Moreton Bay Regional Council n.d.). Timber production continues today on lands managed by the state government and Forestry Plantation Queensland. Agricultural production in Woodford-D'Aguilar today focuses largely on cattle with only two sheep farms recorded. Other key agricultural commodities included orchards and vegetables farms. Mount Mee is also known for its coffee growing and cheese production (Tourism and Events Queensland n.d.).

18.1.2 Wild Dog Management Context

Institutional Context

Parties with an interest in wild dog management in the Mount Mee region include:

- local rural and peri-urban landholders;
- the Moreton Bay Regional Council (MBRC);
- the Queensland Parks and Wildlife Service (QPWS), as managers of the D'Aguilar National Park;
- AgForce Queensland, a farmer representative organisation, which has previously taken on a lobbying role for improved wild dog management and support (Australian Broadcasting Corporation 2010).
- the National Wild Dog Facilitator (assisting local stakeholders with wild dog management planning); and
- the Mount Mee Wild Dog Working Group.

Relevant Legislation and Policy

Wild dogs are a Class 2 declared pest animal under the Queensland *Land Protection (Pest and Stock Route Management) Act 2002* (The State of Queensland 2002). Class 2 pests are generally widely established throughout Queensland, but landholders are required to take reasonable steps to control these species given their ongoing threat to biodiversity, health and safety, or production. Management seeks to limit their impact through population reduction (Moreton Bay Regional Council n.d.).

The Land Protection (Pest and Stock Route Management) Act also requires that every local government in Queensland establish a local government area pest management plan, to "bring together all sectors of the local community to manage pests… ensur[ing that] resources are targeted at the highest priority pest management activities, and those most likely to succeed" (Queensland Government n.d).

While the onus is on the private landholder to reasonably manage pest animals such as wild dogs on their land, Moreton Bay Regional Council acknowledges the difficulties in management for many landholders. It therefore seeks to work cooperatively with landholders through its Pest Animal Management Officers by providing services including trapping, shooting, baiting and pest management plans (Moreton Bay Regional Council n.d.). Council has supplied 1080 bait to farmers, and facilitated coordinated baiting programs with neighbouring farmers (Moreton Bay Regional Council 2010). The Council offers information on wild dog management, and collects data on wild dog attacks in the Council jurisdiction (Moreton Bay Regional Council n.d.). The Council's dog baiting, trapping and shooting program costs approximately $600,000 annually.

The QPWS "Good Neighbour Policy" guides the management of pest species such as wild dogs within natural reserves such as the D'Aguilar National Park (Queensland Government 2013a, b):

> QPWS will seek to co-operate in joint pest control programs with surrounding landholders, other government departments and local governments and will give priority to co-operating with neighbours in the prevention and/or eradication of new outbreaks of pest species... Where a need has been identified, QPWS will also consider approval of 1080 baiting on its lands by lessees, permittees and neighbours, subject to conditions.

QPWS's "Management of Wild Dogs on QPWS Estate" policy spells out approved wild dog control measures within protected lands, responsible parties and the importance of working cooperatively with neighbours and other relevant government agencies (Queensland Government 2015). Since 2010, QPWS has participated in most of the coordinated wild dog baiting programs undertaken in the Mount Mee district, where Park boundaries coincide with private rural landholders. However, because large parts of the Park boundaries are shared with State Forests or privately run plantations, large sections of the Park are not baited.

Dingoes are defined as both "wildlife" and "native wildlife" under the *Nature Conservation Act 1992* (The State of Queensland 1992). This means that dingoes are protected within National Parks and similar nature conservation areas but have no legal protection outside of these protected areas and state forests (Queensland Government 2016). Therefore, within the D'Aguilar National Park dingo populations are protected as they are considered to provide some control of foxes, feral cats and other feral animal species (Queensland Government 2013a, b).

The Queensland Department of National Parks, Sport and Racing (NPSR) is currently responsible for management of the D'Aguilar National Park. A 2013 management statement for the Park suggests that both wild dog and dingo populations present near the Park boundaries are controlled by Department staff, working with landholders neighbouring the Park, as well as with Biosecurity Queensland and the MBRC. Control activities are undertaken several times per year (Queensland Government 2013a, b).

Management Activity

2009 was a particularly difficult year for local farmers, who faced heavy losses of livestock as a result of wild dog populations (Moreton Bay Regional Council 2010). A public meeting led to the development of the Mount Mee Wild Dog Working Group. The group consisted of representatives from the Council, the National Wild Dog Coordinator, NPSR, Forestry and a community member. The group's goals were to discuss and oversee implementation of "nil-tenure" control efforts such as baiting programs (Moreton Bay Regional Council 2010), with the aim of providing "a coordinated and cooperative wild dog management program to reduce the impacts of wild dog predation in the Mt Mee and surrounding areas" (Moreton Bay Regional Council 2010).

Working with local and relevant agencies, the group developed a management plan for Mount Mee and the surrounding areas of Campbells Pocket, Rocksberg, Bellthorpe and Stony Creek. The plan included a goal of aligning these locations into "group areas" for a more strategic and coordinated approach to wild dog management (Moreton Bay Regional Council 2010).

Coordinated wild dog baiting was undertaken twice in 2010, growing to four times per year from 2011. Between 2010 and 2011, landholder participation in coordinated baiting increased by 26% (Moreton Bay Regional Council 2010).

Baits are supplied by the Queensland Department of Agriculture and Fisheries, and distributed to landholders who have signed up for the coordinated activity, at special MBRC-staffed "baiting stations" set up across the Council region. Some 1.2 tonnes of bait are now used annually across the MBRC jurisdiction. However baiting programs using 1080 are becoming more difficult to implement due to demographic changes in the district, and a gradual shift from larger scale production to smaller peri-urban farming and lifestyle land use (Moreton Bay Regional Council 2010).

The more intense focus on wild dog management in and around Mount Mee appeared to have resulted in a 99% reduction in the reporting of livestock predation from 2009 to early 2011 (Moreton Bay Regional Council 2010).

18.2 "The Community Won't Be Ignored": A Narrative of Local Government and Community Action Compiled from 12 In-Depth Interviews

Back in the 1990s, careless landholders sabotaged a community-led baiting program that had started up on Mount Mee. Baits weren't being properly laid out or collected. It got sloppy and there were accusations that some bait had been used to deliberately harm domestic animals. Some influential and knowledgeable landholders walked away from the community-led program, even though it had been pretty successful. There was no trust between members of the community to do the right thing.

When the state government started delegating responsibility for pest animal management down to the local government level, there was a period when wild dog control was not coordinated by any agency. Some of the larger landholders continued to get together to try and coordinate their baiting. Those landholders with a long family history in the neighbourhood had always worked together to manage the country, keeping grass and invasive species under control. When landholders started to notice dog attacks, they would get in touch with one particular neighbour who was getting people on board with some coordinated action. He saw the need to work together and although it wasn't a formal group at this stage, they began to coordinate efforts to meet their landholder responsibilities.

There were a lot of hassles involved with meeting the wild dog control regulations, such as notifying neighbouring properties when a baiting event was planned. The requirements meant hours driving around the mountain, leaving letters for landholders to sign. Sometimes they wouldn't find the letter or they'd forget to leave it in the letterbox for pick up. Getting the baits was difficult too. The group would have to order them from the abattoir and then someone would have to wait around to pick them up. It ate up time and a lot of people just couldn't be bothered. At one stage, the group managed to get about eight or nine people involved in the baiting but slowly they dropped away because it was just too difficult. Some of the landholders then got together and set up the "Mount Mee Cooperative Dog Fund" to pay a couple of shooters to come and knock back the numbers. The group would pool their money and pay 50 bucks for every dog that was shot, but after a while, it became too expensive to continue. This fund then became the seed of the Mount Mee Wild Dog Working Group.

In 2009, the media drew attention to an outbreak of wild dog activity in the Mount Mee area, and the impact on landholders who were losing cattle. This media attention was the first time the Moreton Bay Regional Council ("Council") heard about the problem because there was no effective reporting system in place and so landholders hadn't been able to report increased wild dog sightings or impacts. Council started to cop some flack from the public about wild dog control. The issue of wild dog management started to get political. The Council Mayor and elected members wanted to put political pressure on the state government, because the message they were getting from the community was all about state lands not implementing any effective wild dog management.

There is no doubt that the community lumps all forms of public land together when it comes to wild dog problems. They see it all as "National Parks", a place where management is not effective and a source of wild dogs which then come out and impact on private landholders. But even public lands have different owners—the D'Aguilar National Park ("the Park") shares boundaries with state forest in this case. And over time, some parts of the forest have been privatised, or converted into national park, so the boundaries keep changing. Landholders who back onto state forest lands are not sure if the forestry managers ever bait or shoot wild dogs, and they fear that this gives wild dogs somewhere to hide and breed. As a result, the Parks staff have had some bad experiences with landholders, people getting cranky with them and running them off the mountain.

Darren Shiell was the Council Officer who took on the task of resolving this problem. He could see that Council needed to show leadership and got in touch with Greg Mifsud, the National Wild Dog Facilitator (NWDF) to ask for some help. The NWDF is independent from local government and is guided by the National Wild Dog Action Plan which recommends collective action strategies for wild dog control. Greg took up Darren's invitation to get involved and began by proposing a strategy to get everyone sitting around the same table. A meeting was held up at Mount Mee. Darren and Greg did a lot of work with the leading landholders to get them to listen; Greg also got agreement to have one National Park Officer at each table of landholders. This was a strategy for breaking down barriers and encouraging each party to listen and share information. The Park officers were able to listen to the landholders and see that they were serious about the issue. Based on the advice of the NWDF, Council decided not to implement a bounty program but to try a nil-tenure approach[1] to setting up a coordinated control program. The relevant state minister was asked to commit to trialling this approach. Getting this commitment from as high up the chain of command as possible gives operational staff permission to work collaboratively, which is really important. The Parks staff have since realised that silence is not your friend when dealing with the community because often they will complain to the minister and then everyone cops it. The community won't be ignored.

Darren approached the issue as problem to be solved. Under his leadership, Council has led the development of a range of strategies that made it easier for landholders to be part of a regularly coordinated control program. The first problem was landholders not reporting the issue. Now online reporting is standard across the entire Council area and there is a systematic response to these reports, with Council generating property maps and baiting maps for landholders to look at.

Using maps is one of the key strategies in the nil-tenure approach to planning. It gives everyone something to look at, and when you take the land boundaries off, it is easier to think about how the dogs are moving across the whole landscape—it takes the focus off the individual. A lot of farmers have good information about what is actually happening. Using the maps is also a good way to show the links between public and private lands. When the landholders accept those links and the need to work together, that's a real icebreaker. Based on this information, the landholders are then encouraged to participate in a regular baiting program.

The baiting program started off small but the coordination effort had a positive impact on dog numbers almost straight away. Darren kept an eye on participation and impact, and now the baiting schedule runs four times a year in response to dog breeding cycles and cattle growing seasons. The baiting program uses signage to spread awareness because while the legal responsibility lies with the landholder, not all of them are able to effectively coordinate their control efforts. Sometimes, because of the regulations, landholders weren't able to bait, so the Council program tried to fill in the gaps with trapping. That can be really successful. A big advantage is being able to show a dead dog at the end of it!

Making pest control easier to coordinate is the name of the game. Council sends out the information and pays for the baits. Signage on the side of the road lets people know when baiting is coming up, so you don't have to rely on letterboxes, and that is seen as a great service to the community. The introduction of safe and easy procedures have minimised landholder contact with the poison and this has encouraged people with young children or concerns about health risks from exposure to get involved. It has been getting easier and easier to be involved in coordinated baiting.

Without Council and state government support, it wouldn't be possible to achieve good wild dog control, because the community is not able to do certain control measures due to the regulations around baiting or shooting near boundaries. Since the Council came onboard they have done a great job of breaking down barriers for

---

[1]Nil-tenure is defined in the Glossary.

landholders. They've reduced the paperwork, improved the supply and quality of baits and reduced the costs. The result has been a drop in the number of wild dogs reported and less cattle attacks. While those landholders who have been involved for a long time would have kept trying to coordinate the baiting as part of their civic duty, they are happy that the Council stepped in and took control of the program. There is a lot of rigmarole involved in laying baits or planning shoots, and this all takes time and effort to organise. Now the Council takes charge and landholders don't need to know the details of how they do it. It makes it easy to do the right thing. Having the support of elected Council members is important too, as they are visible in the community.

Mount Mee presents special challenges. The increase of small land holdings is making it harder and harder to get coordinated control. Small properties and tree-changers just aren't interested in dog control—they might not have cattle to worry about, or they work in town and don't know what's going on their block during the day. It's difficult because unless the community reports their pest issues, the problem doesn't exist. At the same time, the peri-urban context is actually the best for nil-tenure, because there are lots of interested parties with a real need to work together. The community needs to know about the ecology of wild dogs, so the problem can be identified. Even for landholders who don't run any livestock, the evidence of wild dog attacks like cow and calf carcasses are distressing to see, and this is usually enough to convince them to join in attempts to control wild dogs.

It would be good if there were a more ethical or humane way to control the dogs than 1080—this would help to increase acceptance of baiting in peri-urban areas. People don't like 1080. They don't like the way it kills the animal and think it's cruel. Some of the National Parks staff don't like 1080 either. They worry about the bi-catch and the ecological impact of baiting and trapping on the pack structure. There have been some sad stories, with domestic dogs being poisoned by accident. It would be great to get some better baiting tools, like the new injectors, so that people could at least find the dead dogs, rather than not see any evidence of the success of the bait and be asked to take a leap of faith. Baiting has a real image problem because people can't see the dead dogs, so they think it doesn't work. Or they see a family pet take bait and they get upset. If they could see the impact that the baiting was having on dog numbers, they might not be so anti-baiting.

The bigger landholders are committed to supporting the Council efforts, so they don't walk away from the program. They agree that the best approach is where landholders cooperate with the Council workers to support the baiting and trapping programs. Under Darren's leadership, the Council staff show respect to the community and try to avoid using academic or overly technical language. They do a lot of one-on-one visits with landholders, sitting around the table, responding to their concerns. The same stories pop up in wild dog management all around Australia— like when shooters claim that baits don't work; or that people won't get onboard because of the myths about 1080. The National Wild Dog Action Plan approach is to bring knowledge to the table and try to dispel the myths. Council officers take the same approach, responding to reports from landholders by sharing information from the area. Sometimes they might instal motion sensor cameras so they can have a look together and identify whether they are dealing with wild or domestic dogs. The landholder's knowledge is really useful for planning the program, and once they have a chance to get past the emotional stuff, they will usually tell you all they can.

In areas where people don't have much experience, they need expertise and action to make things happen. This is the job of the local government, the intermediary between the general public and state institutions. The Council officers really have to know their stuff because people are ringing up to ask questions, particularly when they see the 1080 baiting signs on the side of the road. There's community opposition too, the signs get pulled down or covered in graffiti. The Council officers just keep working away on the program, being open and transparent, and always ready to answer questions.

It would be good if more landholders would get involved in the activity and there is really no reason for not being involved now that the Council has simplified the process. Lazy landholders are a real problem, but sometimes it might be that city folk moving to the country don't understand what they need to do. New landholders are asked to join the coordinated program by their neighbours and the Council officers follow them up. Council has also been successful in reaching out to respected and knowledgeable landholders who may be influential in their community. While landholders don't want to lecture or nag their neighbours, they do take responsibility for reminding people about baiting events and following up if they don't turn up on baiting day. After all, baiting day is a good chance to catch up with other landholders in the valley—have a cup of tea and a chat, share information. It helps build confidence for those landholders who aren't experienced with baiting or trapping. Being part of the coordinated program also means landholders can be safe in the knowledge that all the regulations are being met. Word of mouth is really important for getting people involved and at Mount Mee, the landholders have worked together to implement a "good neighbour" policy, which encourages people to be involved and support the local community action.

Solving these problems is challenging and labour intensive, but the result is an efficient, centralised reporting system that can coordinate pest management and identify gaps in the landscape. The program has knocked back the dog numbers and as long as numbers are kept down in the single digits, the Mount Mee landholders see the baiting program as a success. The program has continued for 6 years and evolved in response to the local conditions. Reducing the administrative load on landholders has made it easier to be involved. The use of nil-tenure planning has also been important because it's not just a once-off intervention—it's about making a plan and getting commitment from all the different stakeholders. It's a proactive response to a problem. Nil-tenure is a just a process—but it's the best one for getting individuals to think about the problem at a landscape scale. And for getting everyone involved and committed to action.

Although the number of attacks has dropped away, it's important that the landholders understand that they have to keep on baiting. The Council program has been successful, but the problems never completely go away: there are still gaps in the program's coverage of the landscape; landholders still don't report their dog numbers; and some of them stop participating once the numbers have been knocked back.

Unless you keep up the pressure, the dogs will re-establish, because they are smart animals that learn how to get around the landscape. For the Council team, it's hard work but really rewarding. They love getting out there, meeting people, seeing the country and helping to keep the feral animals under control.

18.3   Narrative Analysis—Mount Mee

The story of the Mount Mee wild dog management program makes an unlikely start with the description of a failed attempt to implement community-led action. This early attempt by farmers to coordinate a community baiting program was undermined by concerns about the safety of the baits and fears of impacts on domestic animals. There was a breakdown of trust in landholder relationships, and a loss of community leadership. Wild dog impacts continued to exert pressure on farmers at Mount Mee and community leaders continued to look for community-led solutions. Barriers to sustained participation were high. Landholders invested substantial personal time in meeting the baiting regulations, which became too difficult to maintain. They also found it difficult to sustain the financial costs of paying a wild dog bounty and had to close this program. However, these continued efforts demonstrated a willingness for landholders to work together to address a common problem. It is likely that this long history of community-led activity was crucial in the subsequent success of the Moreton Bay Regional Council ("Council") coordinated wild dog control program.

In this case, we see some of the common challenges facing wild dog management in Australia. In particular:

- the role of the media in generating public attention and political action;
- the lack of trust and poor working relationships between private and public landholders; and
- limited information about the number of wild dogs in the landscape.

These challenges create highly charged situations where anger, fear and blame can dominate community interactions. Despite legal responsibility for wild dog management sitting with local government and the affected community, in this case, local politicians reacted to increasing public concern about the impacts of wild dogs by attempting to deflect responsibility to the state government.

At this crucial point, a Council staff member reframed the issue as a problem of community action and Council coordination. This is an important moment in the story of Mount Mee. By accepting the localised nature of the problem, the Council prevented a top-down, state-imposed response. The scale of the problem was clearly identified as *local*, and requiring a local response. There was no promise of an external saviour but rather a commitment to work with the local community to implement a collective action response. The history of community-led attempts in the area created a pre-existing network of community capacity and leadership that the Council could plug into. This then became the backbone of the Council's wild dog program design.

Community concern can be a catalyst for getting people to the table to discuss an issue. However in a highly charged situation, this can often lead to further conflict or increased tensions. The Council invited the National Wild Dog Facilitator (NWDF) to help formulate a strategy to bring the necessary people and interests into the discussion. As an independent observer who does not have a vested interest beyond enabling wild dog control, the NWDF draws on previous experience from around the country to build understanding and break down tensions. In this case, the NWDF brokered political support from the relevant state minister that enabled public lands staff to work collaboratively with the private landholders. This was a breakthrough in historic private–public land management tensions, and gave the government's staff permission to consider how private landholders might see the issue. Through this interaction, it became possible for both production and conservation values to find a place at the community action table.

A strategy of nil-tenure planning, as outlined in the National Wild Dog Action Plan, was recommended as the best way to share information and break down communication barriers. The planning and preparation for the eventual meeting of concerned individuals was extensive, and provided the Council with support to get the best possible result from the planning meeting. Communication was facilitated by looking at maps and combining local landholder knowledge with scientific data. The movement of dogs across the landscape was made visible, regardless of property boundaries, and this use of maps has since become one of the key engagement strategies for the Council in their ongoing program.

The result of the meeting was a commitment to a coordinated baiting program. Council's strategy was to lower as many barriers to participation as possible. This included Council coordination of the baiting schedule; provision of baits, signage and record keeping; regular updates of control maps; efforts to contact new residents; and where necessary, filling in gaps with a dog trapping program.

Importantly, the Council's willingness to drive the program coordination *did not* replace community action. Community service became part of the motivation for individual landholders to participate in the collective baiting program. Community members and Council staff alike saw a public good in addressing a shared concern. Local government politicians supported this by articulating a sense of civic duty that resonated with the community. A persuasive and supportive atmosphere for participation was created. Key community leaders supported the program and actively participated in not just the baiting, but the development of a "good neighbour" policy based on building trust, sharing information and driving a norm of participation in coordinated control. As the demographics of the area keep changing, these attempts to build relationships is a crucial part of the program's ongoing success. New landholders may not have the same values or interests, but may be receptive to a message of shared impact. The success of this message depends on how it balances self-interest with community wellbeing.

In this case, we see an exciting example of how sensitive and reflective coordination can foster a growing sense of civic duty and shared impact across a range of landholders. Council staff and the NWDF see their role as hard work but rewarding, because they are making an important contribution to the economic, emotional and social well-being of the community. Council has provided coordination that allows the community to build on its strengths and take the lead on motivating collective action. The Mount Mee Wild Dog Program fostered a hybrid form of community-led action that has continued to grow over the past 6–7 years.

References

Australian Broadcasting Corporation. (2010, February 14). Gone to the dogs. Television series episode. *Landline*. http://www.abc.net.au/landline/content/2010/s2819075.htm. Accessed February 15, 2016.

Branco, J. (2013, September 25). Mt Mee farming: doing council's work" wants new system. *Sunshine Coast Daily*. http://www.sunshinecoastdaily.com.au/news/mt-mee-farmer-doing-councils-work-wants-new-system/2031890/. Newspaper article. Accessed February 15, 2016.

Moreton Bay Regional Council. (n.d.). *Caboolture township history*. https://www.moretonbay.qld.gov.au/general.aspx?id=348. Accessed February 15, 2016.

Moreton Bay Regional Council. (n.d.). *Pest animals*. https://www.moretonbay.qld.gov.au/general.aspx?id=19435. Accessed February 15, 2016.

Moreton Bay Regional Council. (n.d.). *Wild dogs*. https://www.moretonbay.qld.gov.au/general.aspx?id=9177. Accessed February 15, 2016.

Moreton Bay Regional Council. (2010). *Wild Dogs—Mt Mee: Lifestyle & Amenity Committee Session Report (A2002094)*.

Moreton Bay Regional Council. (n.d.). *Map of council divisions*. https://www.moretonbay.qld.gov.au/maps/. Accessed December 19, 2017.

Mt Mee and Surrounding Area Cooperative Wild Dog Management Plan 2010–2015: Proposed Draft.

The State of Queensland. (1992). Nature Conservation Act 1992. *Legislation*. https://www.legislation.qld.gov.au/LEGISLTN/CURRENT/N/NatureConA92.pdf. Accessed February 15, 2016.

The State of Queensland. (2002). Land Protection (Pest and Stock Route Management) Act 2002. *Legislation*. https://www.legislation.qld.gov.au/LEGISLTN/CURRENT/L/LandPrPSRMA02.pdf. Accessed February 15, 2016.

Tourism and Events Queensland. (n.d.). *Mount Mee*. http://www.queensland.com/destination%20information/mount-mee. Accessed February 15, 2016.

Queensland Government. (n.d.). *Local government pest management plans*. https://www.daf.qld.gov.au/plants/weeds-pest-animals-ants/pest-management-planning/index-to-plans/local-government-area-pmps. Accessed February 15, 2016.

Queensland Government. (2013a). *D'Aguilar National Park, D'Aguilar National Park (Recovery) and Byron Creek Conservation Park Management Statement 2013*. Department of National Parks, Recreation, Sport and Racing. https://www.npsr.qld.gov.au/managing/plans-strategies/statements/daguilar.html. Accessed December 19, 2017

Queensland Government. (2013b). Good Neighbour Policy. *Policy document*. Department of National Parks, Sport and Racing. http://www.nprsr.qld.gov.au/policies/pdf/op-pk-cor-good-neighbour-policy.pdf. Accessed February 15, 2016.

Queensland Government. (2015). Management of Wild Dogs on QPWS Estate. *Policy document*. Department of National Parks, Sport and Racing. http://www.nprsr.qld.gov.au/policies/pdf/op-pk-nrm-wild-dog-mgt.pdf. Accessed February 15, 2016.

Queensland Government. (2016). *Wild dog control and the law*. https://www.business.qld.gov.au/industry/agriculture/land-management/health-pests-weeds-diseases/pests/wild-dog-control/wild-dog-control-and-the-law. Accessed February 15, 2016.

Chapter 19
Case Study: Ensay and Swifts Creek Wild Dog Groups—East Gippsland, Victoria

Abstract The neighbouring communities of Ensay and Swifts Creek face similar threats from wild dogs but are shown to respond in very different ways. This case study reveals how local context influences the way an issue is understood by different communities. Through dissent and discord, a significant connection between knowledge of the issue and power to steer the agenda is identified.

Despite their differences of opinion, each group articulates a common goal to sustain a viable sheep industry. They share similar concerns for the survival of their townships and a desire to actively manage their own destiny. It is in their approach to gaining new knowledge and working with government that their paths begin to diverge. Each develops a strategy that draws on the perceived strengths of their landholder community. This reveals how uneven development of education, expertise and political capabilities can be influential in shaping a community response.

Information, knowledge and recognition all emerge as significant factors in developing a confident community-led program. These cases document a new phase of government investment in community engagement for wild dog control. There is an indication that awareness of how important context can become part of a new engagement practice in the state. It is likely that the benefits will extend beyond the sheep industry to the capacity of the local community.

19.1 Case Study Context

19.1.1 Geographic and Physical Context

Swifts Creek and Ensay are small rural communities located in the East Gippsland region of Victoria, on the Great Alpine Road between the towns of Bairnsdale and Omeo. They are approximately 15 min' driving time from each other, while the

T. M. Howard et al., *Community Pest Management in Practice*,
https://doi.org/10.1007/978-981-13-2742-1_19

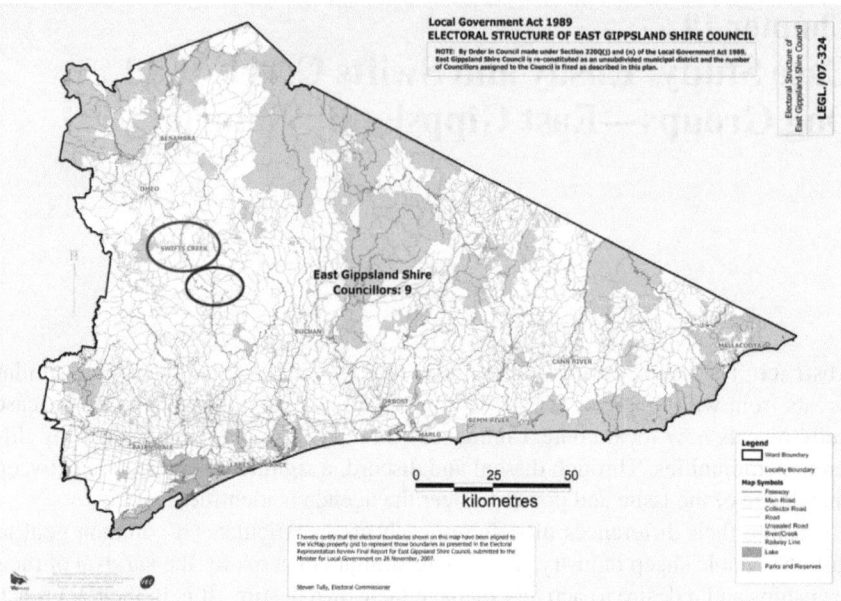

Fig. 19.1 Ensay and Swifts Creek (circled) within the East Gippsland Shire Council (Reproduced from Victorian Electoral Commission n.d.). © Victorian Electoral Commission, made available for reproduction under Creative Commons licence

northernmost community, Swifts Creek, is approximately 70 min' driving time from Bairnsdale. Both towns are stops for tourists travelling to Omeo and further to the Victorian Alps (Fig. 19.1).

Ensay sits near the confluence of the Tambo and Little Rivers. European settlement began in 1843. The area became a soldier settlement site in the years after World War I, and supported a primary school, two pubs, two churches, a community hall, bush nursing service, and a post office. However, in recent decades the population of Ensay has contracted, and some of these services are no longer available in the town (Wikipedia 2014).

Swifts Creek sits at the confluence of Swifts Creek and the Tambo River, and was opened up to European settlement in the mid-1800s during the gold rush period. A key industry in the town today is a sawmill that specialises in making wood pallets. Local services include a hotel, general store, gallery, bookshop, cafe, post office, a caravan park, and nearby tourist cottages (Wikipedia 2015).

Data (Australian Bureau of Statistics 2010–2011) on the production of agricultural commodities in 2010–2011 for the broader region of Bruthen-Omeo shows that production focuses largely on cattle (375 agricultural businesses holding a total of just under 80,000 head of cattle). Sheep grazing is also common, with 169 agricultural businesses holding some 176,062 sheep in 2010–2011.

Ensay Station was once renowned to have the busiest shearing shed in East Gippsland, highlighting the focus of early agricultural production on sheep grazing (Little River Inn n.d.). Today, cattle and sheep grazing remain the key focus of agricultural production in the Ensay district (Ensay Winery n.d.).

Historical agricultural production in Swifts Creek included dairy cattle which serviced a butter factory operating in the town from 1907 until 1946. In addition to plantation timber and the currently operational sawmill, cattle and sheep production are the most notable modern primary industries in and around Swifts Creek (Wikipedia 2015).

19.1.2   Wild Dog Management Context

Institutional Context
Parties with an interest in wild dog management in the Ensay and Swifts Creek districts include

- Local rural landholders;
- The Swifts Creek-Ensay Landcare Group;
- Gippsland and North East Wild Dog Management Groups;
- Bring Sheep Back to Ensay Group (which evolved into the Australian Wool Innovation (AWI) Best Wool/Best Lamb Group);
- Community Wild Dog Control (CWDC): participant landholders in Ensay and Swifts Creek (State Government of Victoria 2015);
- The National Wild Dog Facilitator;
- Victorian Department of Environment, Land, Water and Planning; and
- Parks Victoria.

The East Gippsland Shire Council does not appear to have a role in wild dog management, with these functions instead taken on by relevant Victorian Government departments.

Both Ensay and Swifts Creek have wild dog control groups. Funding administration for both groups is provided by the local Swifts Creek and Ensay Landcare Group Inc., AWI initially funded the groups, with some involvement from the Victorian Government. AWI continue to provide funding to both groups for on-ground control activity.

Relevant Legislation and Policy
Wild dogs as well as dingo-dog hybrids are declared an "established invasive animal" on all public and private land tenures, under the Victorian *Catchment and Land Protection (CaLP) Act (1994)*. Under this Act, all landowners (Agriculture Victoria 2015a, b):

have a legal responsibility to control declared established pest animals under section 20(1)(f) of the Catchment and Land Protection Act (1994), "In relation to his or her land a land owner must take all reasonable steps to prevent the spread of, and as far as possible eradicate, established pest animals."

The Victorian Government makes information available to landowners to explain their legal obligations with regards to invasive animal management, as well as information on the various parties responsible for control in Victoria. At the time of writing (2016), the government's roles for managing invasive species were primarily implemented by the Department of Economic Development, Jobs, Transport and Resources (DEDJTR) and Department of Sustainability and Environment (DSE), with the management of parks and reserves directed through Parks Victoria. The Victorian Catchment Management Council (VCMC) and Natural Resource and Catchment Authorities (NRCAs) undertake advisory and coordinating roles in support of the Victorian Government. Under the CaLP Act, the state government is required to take reasonable steps to control restricted pest animals on any land in the state, including wild dogs. Local governments are responsible for managing established pest animals such as wild dogs on land they manage (Agriculture Victoria 2015a, b).

Dingoes are listed as a threatened species under the Victorian *Flora and Fauna Guarantee Act 1988,* and protected under the Victorian *Wildlife Act 1975.* As such, it is an offence to kill dingoes without authorization to do so. Nonetheless, in 2010 a change was made to the Wildlife Act, declaring the dingo as unprotected wildlife in certain areas of the state in order to allow continued control of wild dogs and (often indistinguishable) wild dogs where they pose a threat to livestock. In 2013, this change to the Act was extended for a further five years (Department of Environment, Land, Water and Planning Victoria 2015).

The Victorian Government has recently implemented a five-year action plan for wild dog management (2014–2019). In addition to improving the management of wild dogs and reducing their impact, a significant goal of this plan is to strengthen coordinated wild dog management across the state, as well as to promote community-based and local approaches to wild dog management as part of the Victorian Government Community Wild Dog Control Programs (CWDC) (State Government of Victoria 2015). This includes working with local communities to develop regional "wild dog management zone work plans", which amongst other things will seek to foster community group formation and participation, and explore opportunities to coordinate public and private land manager efforts. Nil-tenure approaches are favoured for their effectiveness (State Government of Victoria 2013).

As part of the state's five-year wild dog action plan, a wild dog management zone work plan for 2015–2016 has been developed by the Victorian Government for Omeo, Swifts Creek, Ensay and Benambra, following public consultation held in all four locations (State Government of Victoria 2013). The zone work plan includes details on recommended control practices, responsibilities of land managers, participation in community wild dog programs, local contact details, and planned activities for this period, including coordinated baiting programs (State Government of Victoria 2015).

Management Activity

The Victorian Government has increasingly sought to capitalise on local volunteer efforts to control wild dogs with an increased focus on facilitating coordinated activity (Oldfield 2012). Trials involving private landholders and Victorian Department of Primary Industries (DPI) staff have increased baiting coverage on public land adjoining farms in the Ensay district.

Because of the large and relatively inaccessible public lands in the north-east of Victoria and Gippsland, aerial baiting with 1080 poison has been a favoured approach by Gippsland farmers. In the period 2012–2013, the Victorian Government applied to the Federal Government to implement aerial baiting in the Victorian high country. However, the application was rejected due to concerns over potential impacts on native wildlife, despite aerial baiting continuing over the border in New South Wales. Funds were diverted to additional ground baiting instead, at the recommendation of the Victorian Wild Dog Control Advisory Committee. Farmers sought a more strategic and effective baiting program involving a higher density of baits per kilometre (Mulcahy 2012).

The current zone work plan covering Swifts Creek and Ensay suggests that aerial baiting is possible within six Federal Government approved sites across Victoria, (State Government of Victoria 2015) after having been approved during 2014 (Editor 2014a, b, c). Aerial baiting was undertaken in the high country in October 2015 (Poole 2015). Aerial baiting is undertaken annually, and is considered an effective complementary tactic to ground baiting and other control approaches. Private landholders are provided with free 1080 baits in spring and autumn under the CWDC (Editor 2013a, b).

The wild dog bounty resulted in large numbers of pelts handed in across Victoria. The bounty was halted for a brief period before being reinstated in March 2014 (Editorial 2014a, b, c). The wild dog bounty in Victoria ended on 30 June 2015, though the fox bounty continues to operate (Agriculture Victoria 2016).

19.2 "It's Easy to Feed a Farmer's Frustration": A Narrative of Two Community Responses Compiled from 17 In-Depth Interviews

Around 2010 sheep losses from wild dog attacks were horrific and unsustainable. Dog numbers were at plague proportions. Farms had been getting bigger as people sold up, and there were not as many people living in the landscape anymore, regularly baiting or shooting. The stress was terrible. Despite multiple attempts to restock, wild dogs were defeating farmers who had been trying to get back into sheep after an Ovine Johne's outbreak in the early 2000s and this was having significant economic impact in the region. Farmers started keeping sheep in the house paddock at night. There were reports of sheep with their backsides ripped out, still walking around, half eaten. One morning a farmer woke up and thought the snow on the ground around the house was dead lamb. It was always playing on his mind.

At the time, it appeared 90% of the government wild dog program was based on reactionary trapping. Farmers felt alienated from the baiting program because the wild dog reporting system was flawed and the response from the government seemed inadequate for the size of the problem. There were some heated exchanges between farmers and the government department, which seemed to always have a reason why they could not implement any solutions that the farming community suggested.

In the end, government employees are not responsible for the bottom line of private enterprises, so farmers need to take action for their own interests. Every community needs a driver or change agent to step up and point out that wild dogs are not just a government problem, they are a community problem.

Swifts Creek and Ensay are two different communities of interest. Towns located geographically close together often have old rivalries and different social groups. Back in the 1950s, when wool was king, there were football clubs, tennis clubs, schools and businesses. Even with a small property it was possible to make a living from farming sheep. But that is all changed; now there are less and less people in the community and the towns are struggling to survive. Community cohesion across town boundaries has disappeared.

19.2.1   Ensay

The Ensay group were very proactive in dealing with the wild dog problem. An influential landholder got the "Bring Sheep Back to Ensay" group started, got the farmers together and got them talking. The meetings were informal, maybe down at the pub or at a private house. Some initial funding from AWI (Australian Wool Innovation) was really useful for buying some traps and baits to share between the group. After that, it did not take much to get them going. With just a bit of information and coordination the group was ready to take action. The leader knew how to work the politics involved. He was on the Victorian Wild Dog Control Advisory Committee (WDCAC) and had good contacts. The Ensay group had a number of people like this, who had been in different employment or business positions over the years, gaining experience outside the district and the sheep industry. These farmers were good at talking to the government bureaucrats and politicians on their own level.

The Ensay group believed that the government was not adequately baiting the 3 km buffer zone on Crown land adjacent to private properties. The group worked hard and was the first in Victoria to get permission to bait into the buffer on public land adjoining their properties. At first no one thought it could be done, but the members made it happen, and through the drive and skills of the leadership got the bureaucratic support to make the policy changes necessary. The changes allowed the farmers that were being affected to go ahead and bait the area. The government started to consult with the public on co-design of baiting schedules. Once the group members started baiting into the buffer zone, reported sheep losses dropped dramatically.

The Ensay group saw an opportunity to access funding and coordination support under the AWI BESTWOOL/BESTLAMB program. AWI provided support to bring people together under this banner, to share knowledge, visit different farms, and see how they were handling different situations. The meetings would have presentations about a range of issues like lamb health, nutrition, disease prevention and so on, not just wild dog management. They evolved from a single-issue wild dog control group to a grower group focused on producing better wool and better lambs. In the early days, members would spend half the meeting complaining about "Bloody dogs, they're ruining me!" But over time they began to improve their animal husbandry and farm management practices, and it seemed that the benefits outweighed the losses they were getting from the dogs. The biggest change was that dog control became seen as part of the routine of owning a farm, along with all the tools available for more productive sheep farming.

The Ensay group remained informal, with no formal positions such as treasurer or secretary. The local Landcare office looked after any project funds so there was no need to be legally incorporated. The group met when they needed to, for example, when dog numbers started creeping up or baiting days had to be organised. The farmers stayed in touch and rang each other to pass on information. They are deadly serious about killing dogs. They are in charge of their own destination, and have permission to go out there and do the work.

19.2.2 Swifts Creek

Over at Swifts Creek the farmers also had sheep and were battling to keep them. Ovine Johne's was the start of the problem because people had to get rid of their sheep. It was not possible to sell a whole flock and then go out and buy it back again, because it was no longer there to buy back. They lost years and years of breeding. There was a financial and social impact on the region and Swifts Creek farmers felt disadvantaged because they were not as well established as others in the region. Then the wild dogs seemed to get worse. Swifts Creek farmers thought this was due to more baiting than trapping. They felt like the government was against farmers who wanted to keep trapping. There was a core of individuals who vowed and declared that baiting was a waste of time. According to those individuals, the dogs did not eat the baits and just keep breeding up and eating sheep.

The Swifts Creek community felt alone. They were under stress seeing their livelihood being eroded away. Farmers were in the paddock, sleeping with their sheep, and their anger started to build. They were frustrated by red tape and felt that the people making decisions about dog control knew nothing about wild dogs. There seemed to be just "talk, talk, talk" and no action. People were still getting dead sheep every morning. The government department (the department) was not listening to the people whose sheep were being killed, and the problem was symptomatic of public land management. They did not listen because there are no votes in the country areas.

Swifts Creek farmers suspected that the government just did not want them to run sheep at all, although they did not know why.

The Swifts Creek Wild Dog Control Group (SCWDCG) originally started because they resented the Ensay BESTWOOL/BESTLAMB group which seemed to be receiving a lot of support from AWI. Ensay folks were financially set up. They had higher social standing because of their education and financial situation, and as a result, politicians, bureaucrats and industry representatives considered their opinions to be more important. Swifts Creek sheep farmers paid their dues to AWI and wanted to have a say in how that money was being spent in the region. Swifts Creek farmers disagreed with the claims of how effective baiting was because they observed that Ensay farmers had only small numbers of sheep, and in some cases, had gone out of sheep altogether and were now making money from cattle. That was not an option for farmers at Swifts Creek, because the country was not suitable for cattle. Swifts Creek farmers wanted an equal say, and hoped that their group might get the government to do something about the dog problem. They were focused on getting rid of the dogs, not the sheep. There was no back-up plan.

The Swifts Creek group started off informally, when a landholder would ring around the members or visit the shearing shed. The meetings gave everyone an outlet to have a chat down the pub over a couple of beers. With 10 or 15 people at a meeting, it was hard to get everyone to agree but one thing they did agree on was they wanted the dogs gone. Being a small community, farmers were working 6 days a week and then going to footy club meetings, golf club meetings, Landcare meetings and so on. There was not much time left to put into the wild dog group. All they wanted to do was run sheep but all the programs were focused on baiting, baiting, baiting. The group only used half of the AWI money that they were allocated because they did not want to throw it away by chucking baits out in the paddocks when the trapper knew most dogs walk past the baits. That money was given to AWI as a levy from wool production and the group wanted to try and get something of value back, not just drive around chucking baits out the window. The Swifts Creek farmers did not claim to be experts but they had learned from their fathers and friends about how to catch and shoot dogs. They were not educated people, but they had expertise that they felt was not being heard by AWI or the government and were being left out of the conversation.

19.2.3 Policy Action

Farmers from both Ensay and Swifts Creek were represented on the Wild Dog Control Advisory Committee, delivering two-way communication by feeding information to the department and bringing information back to the community. Ensay representatives were focused on solutions. They started to lobby key people with their ideas, and asked "why can't we make it work?" Whereas Swifts Creek members would threaten "We'll go to the press, to the minister," in a confrontational way. But they began to realise that there was not an open purse from the government for wild dog control because the community also wanted more hospital beds and more roads and more schools. If the farmer wanted to stay farming, then he needed to do something himself. There was a genuine attempt by the department to come up with some better outcomes, and they began to realise that the landholders were serious about a meaningful dialogue.

Once the department started to listen to the producers, it became clear that the groups were asking for help, and solution assistance did not have to be financial. Getting out and talking to people, understanding and empathising with what they were going through, helped the department understand the reasons for their anxieties. That kind of informal communication helped uncover the emotion and depth of feeling on the issue. It also helped the farmers see they were not alone, and that somebody did care what happened to them and their business. Before this time, the relationship between landowners and the department was so poor that farmers were reluctant to communicate about their wild dog problem—they just did their own thing.

Dog Controllers
Field staff can be very influential, so if they are selective in the information they provide or denigrate one method of control, they can taint landholder attitudes. It is easy to feed a farmer's frustration and this can lead to a lot of mistrust about how dog control is planned and implemented. For example, there was a myth about wild dog controllers that unless they had 40 years' experience and were second-generation farmers, they could not go into the bush and trap a dog. But a wild dog controller who is proactive in promoting best practice will get a good result.

Catching dogs and talking to people are the skills that a good dogman needs. In the past, dog controllers were low-skilled and on the lowest government pay scale. The WDCAC wanted to address that issue. The controllers needed to develop their skills to take them from a hunter-gatherer into the twenty-first century, using technology and giving them some career progression. Now they drive around with a tablet, GPSing every bait and entering information into the data collection system. A good dogman should be able to communicate and plan with the farmers in a partnership. He can be a hero in that community. Whereas a controller who wants to be seen as the good guy will tell a few white lies and try to put all the blame onto the government.

One of the government dogmen in the region was not very effective and the Ensay group started to complain. It was not a pleasant process but in the end someone with better people skills was brought into replace him. Permanent government employees are hard to move on but no one should make excuses for the dogmen anymore. This is not the 1950s. This is the era of communication and service. They may be rough bushies, but farmers want to know what is going on and to tell the dogmen and their managers when there is a need for attention.

Government and industry have worked hard to address this. In 2014, the department put on a Community Engagement Officer. At first there were complaints that the wages for that position could have employed two wild dog controllers, but the farmers began to realise that maybe the program can work this way. The position can have a significant impact by addressing some of those engagement issues, talking with individuals who have long held concerns and have been very vocal in the public arena and trying to work through some solutions. It is important to counter the arguments, give individuals the opportunity to ask questions about the research or the government policy. They begin to learn how complicated it is to make a good government policy, and how one decision about dog control can influence everybody's bottom line. It may not be possible for government or industry to solve the problem tomorrow, but they can make a start. It takes time to build trust and it is not something that producers give out easy.

External coordination support from industry or government is really important to getting the work done. That coordination makes sure that the baits are picked up, the paperwork is done and the program meets all the regulations. The evidence suggests that once people understand what the problem is and have the information they need, as well as the ability and the accreditation to take action, they are more than willing to come on board. Farmers can contribute to that; they already exchange news when they meet each other on the fence or out on the road. If someone gets a dog attack, the phone will start ringing. The system needs to be in place so they can report the wild dog activity and use the data to make plans. The data is becoming more and more reliable. Trends have emerged about dog numbers, dog kills, properties affected, and this can be broken down region by region, and locality by locality. So if one locality is doing something a bit different, it is possible to measure the success or failure of that. It is a case of laying baits, using traps and watching the impact.

Freedom from dogs means that farmers have the option to run sheep again and keep their sanity. There is a misbelief that if control programs take out big numbers of wild dogs, that will be the end of the problem—it is important to realise that there will never be zero threat. The dog numbers can be managed and constrained to certain locations but dogs have been around forever and a day, and they need to be managed like every other farming problem. There are definite community benefits from effective wild dog control. It keeps sheep farmers in the area, which keeps shearers in the area and the wool sellers and so on. Even cattle farmers have a vested interest in making sure there are enough kids in the community to keep the school open. In the end, the community has to get behind it because everyone benefits from the business of raising sheep and cattle.

19.3   Narrative Analysis—Ensay and Swifts Creek

In this story, two neighbouring communities describe the impact of wild dogs on their sheep industry, illustrating that there is always more than one way to frame an issue. The way a story is told reveals what individuals see as significant, how success or failure is defined, and how a group of individuals makes sense of their shared experience of wild dog management. Viewed in this way, the stories of Ensay and Swifts Creek demonstrate how historical patterns of social standing and financial advantage can persist over time and have real impact on community capacity and willingness to take action.

The changing circumstance of the sheep industry is the background of these stories. Although wild dogs are recognised as a significant cause of financial and emotional stress, many other factors impact on the viability of the industry in East Gippsland. An outbreak of Ovine Johne's in the early 2000s forced farmers to exterminate or trade their sheep out of the region. Cattle farming became widespread, reducing employment in the district and contributing to the depopulation of the country towns. Smaller farms merged into larger enterprises, resulting in less people dispersed over a bigger area. A region that had prospered in the past now struggled to adapt to the changes being experienced in many other agricultural regions around Australia.

A tension between government management of wild dogs on public land and the needs of the farming community emerged as a major driver in both Ensay and Swifts Creek. Dissatisfaction with the government response was linked to a feeling of powerlessness about the imposition of rules and regulations on the landholders' ability to take action. The way each group responded to this driver had significant flow-on effects in terms of institutional support, political recognition and the social welfare of their communities.

19.3.1   Ensay

Where the landholders accepted responsibility for protecting their own economic interest, there was a commitment to work cooperatively with government and industry in developing a path forward. This was the approach preferred by the Ensay group leadership, which became regarded as proactive and politically astute. Commencing with the community-led "Bring Sheep Back to Ensay" project, the group's leadership was open to receiving information about wild dog management from a range of sources, such as AWI. The group was not wedded to one management method and wanted to make evidence-based decisions about wild dog management. Members had a range of experience and contacts to draw on. This bolstered their confidence in dealing with government and industry.

Confidence is a tangible quality that emerges from personal experience of social relationships. It can suggest a historical access to privilege and positions of advantage. The capacity of the Ensay group to incorporate new knowledge into existing farming practices reflected a confident leadership that did not fear the prospect of change. Seeing an opportunity to access coordination support, new knowledge and funding, the leadership embraced the BESTWOOL/BESTLAMB program. This framed the problem as one of sheep production and animal husbandry, shifting the focus from the single issue of wild dogs to broader concerns of economic viability. A deceptively simple act, this reframing increased the range of options available, and empowered the members to try new strategies rather than stay locked in an emotionally draining battle with the dogs.

The efforts of the Ensay representative on the Wild Dog Control Advisory Committee to gain permission for landholders to bait into the 3 km crown land buffer demonstrated a combination of political awareness and confidence in their ability to propose a solution. This achievement gave landholders a concrete example of success stemming from group participation and the approach of the leadership. The decision gave landholders permission to take direct action, leading to a mutually reinforcing connection between group participation and sheep survival.

These concrete examples of successful group action underpin the narrative of Ensay as a proactive and successful group. They illustrate the link between knowledge and capacity to take action, which emerged as an important dynamic in these stories and as a point of difference between the two groups.

19.3.2   Swifts Creek

Mistrust of government and a feeling of disadvantage heavily influenced the way that Swifts Creek landholders responded to the wild dog threat. Whereas external stakeholders perceived Ensay as being open to negotiation and focused on finding cooperative solutions, Swifts Creek were seen as caught in an entrenched pattern of conflict and confrontation with government.

They saw their community as less financially established than Ensay, less educated and with fewer influential networks as a result. Comparing themselves in this way led to resentment and suspicion about the story of success being presented by Ensay. This seemed to exclude the lived experience of Swifts Creek landholders, who felt less and less visible as Ensay's profile grew. This perspective seemed to disregard the political influence gained by Swifts Creek's links with other powerful associations such as the Mountain Cattlemen's Association. Feeling unheard and invisible, the Swifts Creek community saw their situation as a continuation of historical patterns of privilege that actively discriminated against local knowledge and nonprofessional expertise.

The single-minded focus on trapping as the only effective method of dog management reflected an attempt to regain control of the story. A disgruntled government employee with family roots in the area was the local wild dog controller. He advocated trapping over any alternative method and fostered unrest in the local sheep farming community. In this way, Swifts Creek's rejection of an evidence-driven approach to sheep management became a signal of deeper conflict. The Swifts Creek community felt isolated and powerless. Importantly, they felt left out of decisions about how the wool industry funds were spent.

In an attempt to reassert control of the wild dog management agenda, Swifts Creek privileged local knowledge and expertise at the expense of new information. This enabled strong personalities such as the disgruntled wild dog controller to influence opinions, leading to a simplistic binary of trapping or baiting, rather than an adaptive program of management. A lack of organisation limited the group's ability to take advantage of the policy change allowing them to bait on public lands adjoining their private properties. The group was unable to meet the compliance requirements of the policy change such as maintaining a volunteer register and this reinforced the story of disadvantage and inequity, increasing the group's sense of alienation from government and industry support.

The extent to which these feelings of exclusion were based in real and active discriminatory practices is difficult to assess.

Regardless of the factual base for these claims, it is clear that they were extremely influential in determining how the landholders responded to government regulations and industry attempts at supporting community action.

The determination of a subset of Swifts Creek landholders to engage with government through the WDCAC was regarded with suspicion by the majority of the group. However, this initiative recognised the link between access to good quality information and the landholders' capacity to make evidence-based decisions. By actively participating in the advisory groups, these landholders began to influence change on the ground. Qualifications and skills for wild dog controllers were upgraded so they could support better data collection systems. The intersection of knowledge and power were again revealed as crucial to empowering landholders to make well-informed decisions.

Both Ensay and Swifts Creek groups identified as proudly informal, relying on local networks to communicate and organise meetings when necessary. However, both also accessed varying degrees of external coordination support through Landcare, the Catchment Management Authority (CMA), the Victorian Government, the WDCAC and AWI. Whether this support was recognised or acknowledged depended on the individual perspective, with those feeling alienated and isolated less likely to see this as significant to their situation. While this may be frustrating for those working on the ground with the farmers, it serves as an important reminder that context is important when dealing with difficult problems such as wild dogs.

For the farmers of the East Gippsland region, the economic and social survival of their towns and communities is the motivating force for getting involved in wild dog management. These stories demonstrate how overcoming both perceived and real patterns of power and privilege may be facilitated through the provision of good quality information. Beyond access to information, government and industry can focus coordination support on understanding the different history and experience of each community. These stories show that a willingness to listen is a vital ingredient in good quality community engagement. Then strategies can be tailored for the needs and capacity of the specific community, increasing the possibility that each group will find the best way to take action.

References

Agriculture Victoria. (2015). *Established invasive animals*. http://agriculture.vic.gov.au/agriculture/pests-diseases-and-weeds/pest-animals/invasive-animal-management/established-invasive-animals. Accessed February 17, 2016.
Agriculture Victoria. (2015). *Noxious weed and pest management*. http://agriculture.vic.gov.au/agriculture/farm-management/legal-information-for-victorian-landholders/noxious-weed-and-pest-management. Accessed February 17, 2016.

Agriculture Victoria. (2016). *Fox bounty*. http://agriculture.vic.gov.au/agriculture/pests-diseases-and-weeds/pest-animals/fox-bounty. Accessed February 17, 2016.

Australian Bureau of Statistics. (2010–2011). *Cat. No. 7121.0—Agricultural Commodities, Australia, 2010–11*. Canberra, ACT: Australian Bureau of Statistics.

Department of Environment, Land, Water and Planning Victoria. (2015). *Dingoes in Victoria*. http://www.depi.vic.gov.au/environment-and-wildlife/wildlife/dingoes-in-victoria. Accessed February 17, 2016.

Editorial. (2013a, August 30). Wild dog baiting springs into gear. *Stock & Land*. http://www.stockandland.com.au/story/3556605/wild-dog-baiting-springs-into-gear/. Accessed December 21, 2017.

Editorial. (2013b, October 31). Call to lift aerial baiting ban. *Stock & Land*. http://www.stockandland.com.au/story/3557316/call-to-lift-aerial-baiting-ban/. Accessed December 21, 2017.

Editorial. (2014a, January 13). First approval for aerial baiting bid. *Stock & Land*. http://www.stockandland.com.au/story/3555830/first-approval-for-aerial-baiting-bid/. Accessed December 21, 2017.

Editorial. (2014b, February 26). Local wild dog knowledge sought. *Stock & Land*. http://www.stockandland.com.au/story/3555561/local-wild-dog-knowledge-sought/. Accessed December 21, 2017.

Editorial. (2014c, February 26). Bounty back in business. *Stock & Land*. http://www.stockandland.com.au/story/3555566/bounty-back-in-business/. Accessed December 21, 2017.

Ensay Winery. (n.d.). *Ensay Winery*. http://ensaywinery.com.au. Accessed February 17, 2016.

Little River Inn. (n.d.). *Ensay Station Shearing Woolshed Ruins*. http://littleriverinn.com.au/proddetail.php?prod=ensay-station-woolshed-ruins. Accessed February 17, 2016.

Mulcahy, M. (2012). VFF support for summit on wild dogs. *Border Mail*. http://www.bordermail.com.au/story/108412/vff-support-for-summit-on-wild-dogs/. Accessed December 19, 2017.

Oldfield, J. (2012, November 8). Dogfight needs co-operation. *Stock & Land*. http://www.stockandland.com.au/story/3557930/dogfight-needs-co-operation/. Accessed December 21, 2017.

Poole, L. (2015, September 9). Victoria to begin aerial baiting for wild dogs next month as East Gippsland farmers spend thousands on electric fences. *ABC Rural*. Australian Broadcasting Corporation. http://www.abc.net.au/news/2015-09-09/new-dog-fences/6761140. Accessed February 17, 2016.

State Government of Victoria. (2013). *Action plan for managing wild dogs in Victoria 2014–2019*. East Melbourne, VIC: Department of Environment and Primary Industries.

State Government of Victoria. (2015). *Omeo, Swifts Creek, Ensay and Benambra Wild Dog Management Zone Work Plan 2015–2016*. Melbourne, VIC: Department of Environment, Land, Water and Planning.

Wikipedia. (2014). Ensay, Victoria. Encyclopedia entry. https://en.wikipedia.org/wiki/Ensay,_Victoria. Accessed February 17, 2016.

Wikipedia. (2015). *Swifts Creek. Encyclopedia entry*. https://en.wikipedia.org/wiki/Swifts_Creek. Accessed February 17, 2016.

Chapter 20
Northern Mallee Declared Species Group—Esperance, Western Australia

Abstract The Northern Mallee Declared Species Group (NMDSG) case documents challenges faced by a single-species wild dog action group in a changing policy context. A looming state government reform to the funding and management regime is seen to threaten the group's long-running and well-tested model of community-led action.

Initially formed to protect sheep farming interests, the group has been successful in adapting their message to stay focused on wild dog control, while accessing support from other agricultural industries. A strong and highly visible Chairperson leads with passion and conviction. Members are willing to take action in support of the leader, utilising the media contacts and industry networks to implement a strategy of community-led action that is strongly tied to political advocacy.

The Chairperson remains committed to a long running campaign to extend the wild dog fence. This tenacity inspires loyalty from the group members, and sometimes begrudging admiration from government and industry. Strong leadership, a distinct feature of this case, can be difficult to replace and this raises questions about the long-term sustainability of the group.

In the NMDSG case, the increased responsibilities and workloads that come with formalised group structures are clearly illustrated, raising important questions about how to strike the best balance between community-led action and coordination support.

20.1 Case Study Context

20.1.1 Geographic and Physical Context

Esperance is a coastal town located in the Goldfields-Esperance region of Western Australia, approximately 720 km from Perth on the South Coast Highway. It is a popular tourist destination, offering surfing, scuba diving, swimming, recreational

© Springer Nature Singapore Pte Ltd. 2019
T. M. Howard et al., *Community Pest Management in Practice*,
https://doi.org/10.1007/978-981-13-2742-1_20

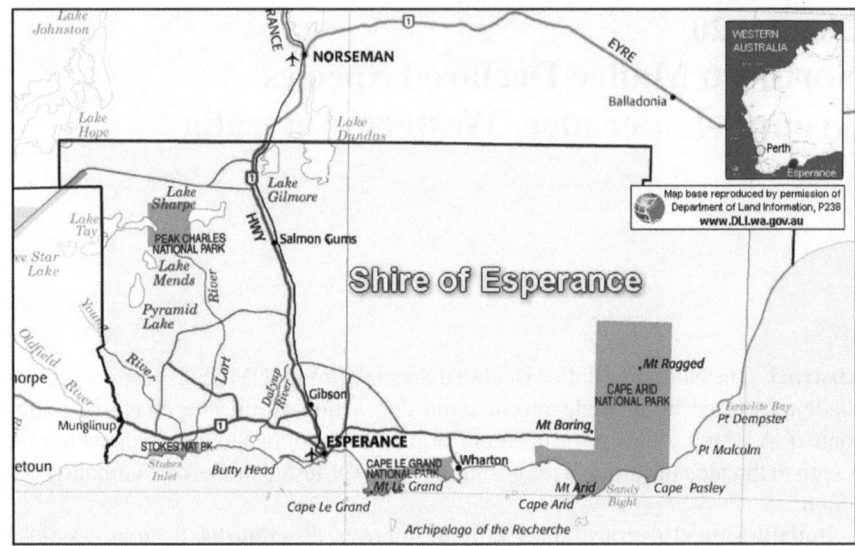

Fig. 20.1 Shire of Esperance, WA, including the town of Esperance and other localities of Gibson and Salmon Gums. (Reproduced from Local Government Network, Australia-LGNet n.d.)

fishing and hiking and four-wheel driving in a number of nearby national parks (Wikipedia 2016). The Shire of Esperance covers some 44,336 km^2 and incorporates over 400 km of coastline (Fig. 20.1).

European settlement of the Esperance district dates back to the 1860s, with the town having a population of approximately 1000 people by the late 1890s. At this time, the discovery of gold in the Goldfields region to the north contributed to rapid expansion of the port town of Esperance (Visit Esperance n.d.).

Agricultural Production[1]

Farming in the Esperance district was slow to establish due to drought, salinity and poor soil quality. However by the 1950s, researchers recognised the potential of the Esperance Mallee country to become viable farming land, with the addition of superphosphate and other soil supplements. Land clearing and agricultural production expanded rapidly over the next two decades, as agriculture became the dominant form of industry in the district (Visit Esperance n.d.).

Combined data (Australian Bureau of Statistics 2011) on the production of agricultural commodities in 2010–2011 for two Statistical Area Level 2 (SA2) regions of "Esperance" and "Esperance Region" shows that they include some 574 agricultural businesses. Total grazing area was estimated at 407,967 ha, of which the majority

---

[1]Detailed agricultural production data are available for the Statistical Area Level 2 (SA2) regions of "Esperance" as well as the broader "Esperance Region", which includes the remainder of Shire of Esperance and most of the neighbouring Shire of Ravensthorpe. These data are not available at smaller geographic scales.

was improved pasture. Just over 1,000,000 ha were devoted to crop production, and an estimated 65,046 ha devoted to forestry.

Agricultural production in the region is a diverse mix of grazing and cropping enterprises. Sheep production is particularly significant, with over 880,000 head being grazed in 2010–2011. Cattle production (almost entirely meat cattle) is also important, with some 120,000 head present in the region. The average amount of grazing area per farm in the Esperance SA2 was 828 ha, while in the larger Esperance Region SA2 it was 1074 ha.

Given the scale of livestock production in Esperance and the surrounding agricultural region, many producers also engage in the production of hay and silage, or cut pasture or cereal crops for hay. More than 950,000 ha is devoted to a variety of cereal and noncereal crops, predominantly wheat, barley, canola, field peas and lupins.

20.1.2   Wild Dog Management Context

Institutional Context
Parties with an interest in wild dog management in the Esperance district include:

- Local rural landholders;
- The Northern Mallee Declared Species Group (NMDSG);
- Shire of Esperance Council;
- South Coast Natural Resource Management (South Coast NRM);
- Pastoralists and Graziers Association of Western Australia;
- Australian Wool Innovation (AWI);
- Department of Agriculture and Food (DAFWA)—became Department of Primary Industries and Regional Development (DPIRD) 2017;
- Department of Parks and Wildlife (DPAW); and
- The National Wild Dog Facilitator.

Relevant Legislation and Policy
In Western Australia, wild dogs, dingo–dog hybrids and dingoes are listed as declared pest species for the whole of the state. Landholders and land occupiers are responsible for declared pest species control on their own properties, under the Western Australian *Biosecurity and Agriculture Management Act 2007*. All three forms of declared wild dog (*Canis lupus* familiaris, *Canis lupus* dingo x *Canis lupus* familiaris, and *Canis dingo*) are considered (Department of Agriculture and Food n.d.):

> Organisms that should have some form of management applied that will alleviate the harmful impact of the organism reduce the numbers or distribution of the organism or prevent or contain the spread of the organism.

Dingoes are considered "unprotected native fauna" under the Western Australian *Wildlife Conservation Act 1950*. This Act allows dingoes to be controlled in agricultural and pastoral regions, though they are generally not controlled in the rest of the state. State government policy is for both wild dogs and dingoes to be controlled in or near livestock grazing areas (Department of Agriculture and Food 2014). Within the national parks, wild dog control is carried out by the Department of Parks and Wildlife (Department of Parks and Wildlife n.d.).

Through the DAFWA, the Western Australian State Government currently offers support for private landholders to control declared pest species via regional, community-based Recognised Biosecurity Groups (RBGs) and Declared Species Groups (DSGs) (Jose 2014). The purpose of these groups is to facilitate coordinated cross-tenure management of declared pests, in support of (rather than as a replacement for) the responsibility of the individual landholder to manage declared pests on their land (Department of Agriculture and Food 2015).

The community-based NMDSG was formed in 2004 and incorporated 2010 (Northern Mallee Declared Species Group Inc. 2014a, b, c, d). The NMDSG has been incorporated since 2010 and employs a part-time administrator as well as two full-time doggers (Northern Mallee Declared Species Group Inc. 2014a, b, c, d). The Group's management structure includes a Chairman and a Secretary/Treasurer. Annual general meetings are held, as well as special general meetings to discuss and vote upon important issues (Northern Mallee Declared Species Group Inc. 2014a, b, c, d; Shire of Esperance 2015). South Coast NRM has provided technical support, and a number of sponsors and corporate members provide financial support (Northern Mallee Declared Species Group Inc. 2014a, b, c, d). The Shire of Esperance handles the Group's finances.

The NMDSG recorded the following achievements for the period 2004–2011 (Northern Mallee Declared Species Group Inc. 2011):

- reduced stock losses;
- extensive targeted hand baiting, working with farmers, better bait placement, expansion of the buffer;
- employment of two full-time doggers;
- successful trial of Maremma guard dogs;
- two coordinated aerial baitings, Autumn and Spring;
- developed and implemented a 5-year management plan and budget;
- direct links to National and State Wild Dog Committees and the State Barrier Fence Committee;
- working closely with stakeholders, Esperance Shire, South Coast NRM, DAFWA and DEC;
- incorporation of the NMDSG; and
- positive media promotion.

The NMDSG became "Esperance Biosecurity Incorporated" in 2017 (Jose 2014).

Management Activity
DAFWA considers baiting to be the most cost-effective method for lethal control of wild dogs, as well as the only practical approach in inaccessible or remote locations (Thompson 2008). The NMDSG has a history of involvement in aerial and ground baiting programs, working with partners such as the Government Natural Resource

Management Program (Natural Resource Management Program n.d.), and South Coast NRM (South Coast Natural Resource Management n.d.). Members of the NMDSG receive 50 free fox baits every year (Northern Mallee Declared Species Group Inc. 2014a, b, c, d).[2]

The NMDSG facilitated the trial of white Maremma guard dogs, bred and trained specifically to protect sheep flocks against wild dog attack (Northern Mallee Declared Species Group Inc. 2014a, b, c, d). In addition to purchasing eight dogs in 2010 to distribute to members for trials, the Group received a donation of a further six dogs from the Australian Dingo Conservation Association (Johnston 2011), however, these trials were not considered a success and were discontinued.

The State Barrier Fence, which is approximately 1200 km in length, has been an important part of Western Australia's strategy to restrict the movement of wild dog populations into agricultural zones. DAFWA staff are responsible for ongoing fence repairs (Department of Agriculture and Food n.d.).

The existing fence does not extend to the coast and as a result, farmers in Esperance are not protected by the structure. The NMDSG and other stakeholders have lobbied the state government to extend the fence to a new termination point near Cape Arid National Park (Northern Mallee Declared Species Group Inc. 2014a, b, c, d). DAFWA developed a proposal to extend the fence around Esperance, and obtained funding from the community as well as the state's Royalties for Regions program, to contribute towards the expected $5,000,000 cost of the fence extension (Department of Agriculture and Food n.d.; WA Country Hour Team 2012). The fence proposal has raised concerns about potential negative implications for native wildlife and natural biodiversity (Driscoll 2013).

Locals continue to work together to implement a range of control measures and to lobby for further support, with the result that stock losses in Esperance due to wild dogs have declined (Northern Mallee Declared Species Group Inc. 2011).

20.2 "Singing from the Same Hymn Sheet": A Narrative of Wild Dog Group Action Compiled from 13 In-depth Interviews

Back in the late 1980s, Esperance had six million sheep and now there's only one million. In the early 2000s, there was a series of big fires out in the crown land north of the Esperance region and these fires made access for wild dogs a lot easier coming down through the bush. At the same time, the Western Australian Government reduced their wild dog baiting programs and as a result, the number of wild dogs exploded. Landholders first noticed attacks on their sheep in the paddocks near

---

[2]The number of baits that NMDSG doggers have laid in the past few years: 2012—12,370; 2013—16,125; 2014—10,951; 2015—9907. Many landholders within the shire lay baits on their properties to help control foxes particularly leading up to lambing but the Group does not collect figures on these (NMDSG Secretariat, pers. comm., May 2016).

unallocated crown land. In a period of 5–6 weeks, one landholder lost 70–90 full-grown ewes worth $100 a head. Out of sheer necessity, and the realisation that no one else was going to do anything, Scott Pickering called a meeting of landholders at Salmon Gums Station in 2004. The main motivation was the impact on his financial bottom line, but also the emotional distress of seeing his sheep being attacked. Wild dog attacks have a big emotional impact on farmers which is often unrecognised. If sheep get worms, you drench them; if they get foot rot, you clip their feet and bathe them; fly strike, you shear them, you crutch them and you treat them. With wild dogs, there isn't much that the individual farmer can do about it. Individuals acting together can do so much more.

It was an informal group at first, but then became incorporated in 2009 as the Northern Mallee Declared Species Group (NMDSG). This helps the group access funding from Department of Agriculture and Food Western Australia (DAFWA) and the Department of Parks and Wildlife (DPAW). The group uses this funding to hire a couple of doggers who lay out baits and set dog traps. The group also supplies members with 50 free baits a year, an incentive that creates a feel-good response in the community too. They see that there is an integrated response to the issue and all stakeholders are contributing to the response.

In the past, government would step in to coordinate species control and that built an expectation about the government's role that is no longer the case. DAFWA capacity has been drastically reduced and a lot of the support they used to offer isn't available anymore, such as helping with funding applications or liaising between farmers about species control. Back in the day, DAFWA had government doggers running around in the landscape. Landholders were more distant from what was going on—all the issues involved in baiting and trapping and accessing different types of lands and dealing with different departments. This created the expectation that pest management was the sole responsibility of the government. That culture of expectation is hard to shift.

The NMDSG has strong, politically astute leadership. A successful group needs a recognised leader, capable of speaking for and representing the interests of others in the group. Scott Pickering is the Group Leader—he's a successful farmer and that makes him credible to start with. He's professional and able to put the business of the group above any personal issues. He has good support from the Treasurer and his wife and the rest of the group leadership team. The group leadership is right on the front line of impact, and they have a direct interest in reducing the number of wild dog attacks on their sheep.

Scott has the knack of confronting politicians or bureaucrats and saying what needs to be said without upsetting too many people. That's a big advantage. He also uses the media well, which helps him define the problem and drive the agenda. Burn out is an issue though. Administration is important for a group to succeed and this skill set can be hard to find in a rural community. Volunteers get worn out, it's the same in all of the farmer groups. Particularly in reporting, applying for funding and keeping up with the governance requirements of being an incorporated body.

Government funding has started to dry up and the grants that do come up are more and more complicated. For a volunteer, running a business and living 120 km from town, to sit down and read the application guidelines and try to understand them is nigh on impossible. Sometimes, it seems as though applications are being written for bureaucrats to fill out. Most farmers learn by doing—they are practical people. They need help with reporting, someone to go over their financials or reports to see if they are on the right track. This can help build confidence within the group, letting them know "yep, you're doing a good job, that's fine". A group might have great leadership but they still need technical advice. They need help to interpret scientific data and government data and transfer it into a language they understand. Otherwise, they look at it and go, "what the hell are they trying to do here? What are they trying to say? This is too scary, we're backing away". The NMSDG has found all this very frustrating and has struggled to get competitive grants. The group just want to go out there, control dogs and protect their livelihoods. There is a need for trusted people who can improve communication between those on the ground and the bureaucrats.

The government has decided to focus resources on aerial baiting, but this is ineffective on its own. It needs to be combined with hand baiting and trapping, which is now coordinated by NMDSG. This has been successful—there are less stock losses and less dogs being caught. The coordination of the different approaches is the reason for the success, and if this coordination stopped, the dog numbers would increase straight away. The group is effective in keeping dog numbers down because they have good doggers working for them and are well organised. All the farmers seem happy, because they're not getting attacks—they soon let you know if they're not happy. Using doggers, baiting and trapping altogether seem to be the only way to keep it down to a manageable level.

If the doggers disappeared, if the group couldn't employ them because they ran out of funding, the dog numbers would jump up and the sheep industry would be gone.

It's heartbreaking to see sheep get mauled by dogs. Some farmers get out of sheep altogether, but even years later they will still be looking for eagles circling, because that's the first sign of a wild dog attack. That's why the Treasurer stays on the NMDSG committee, even though he doesn't run sheep anymore—he doesn't want others feeling that kind of emotional impact. The group says they need a fence to keep the wild dogs out. The proposal to extend the existing state barrier fence to Esperance provides a clear goal for the group and that encourages community and industry support. The group organised a referendum through the local government to see if farmers would pay a levy to finance the fence and this suggestion received good support from the community.

Wild dogs are a serious problem in the Esperance region, but emus also have a big impact on cropping. The group has suggested that the barrier fence can address both of these problems. The Group Leader is passionate about the fence project but is realistic about the delays in the process and is good at keeping the farmers motivated. There are processes to follow in regards to land clearing and Aboriginal heritage, which have slowed the fence project down but it's important to avoid detrimental impacts. Other constraints are native title requirements, threatened species impacts and the risk of tree dieback. The fence extension would segment the Greater Western Woodlands, which is the largest remaining area of intact Mediterranean-climate woodland left on the Earth. Fencing through this woodland has implications for biodiversity and gene flow. DPAW has done a range of flora and fauna surveys to advise the fence development, as required by the application process. The plan for the fence itself has been realigned many times, in response to various pressure points and issues.

Having a single common goal has been part of the group's recipe for success, as well as the Group Leader's ability to get in the ear of every minister he can. Scott has been motivated to find out how the whole system of wild dog management works in Western Australia (WA). He applied to go onto the state Wild Dog Committee, and then the state Barrier Fence Committee, and that led to him being on the National Wild Dog Committee. He's also active in the merino industry body and leverages those connections to benefit the group.

DAFWA established the Esperance Extension Reference Group, which both the NMDSG Group Leader and Treasurer sit on. Coordination services are funded by a WA government Royalties for Regions (R4R) grant. The group argues that the Esperance region is being unfairly disadvantaged by the reduction in government support through DAFWA because it was only opened up for agriculture relatively recently, unlike other areas of the state. Another R4R project provides about $90,000 per year to the group to pay for a dogger. DPAW runs regular aerial baiting on their lands throughout the year and apportions some of their feral animal management budget to support the NMDSG dogging program. The dogger uses a data logger to keep track of baits and traps and this is part of the reporting back to the funding body. Industry pressure from both the grain and small livestock industry has helped the group get this funding support.

Part of the group's success is that everyone on the committee is singing from the same hymn sheet. The group will usually get together before a meeting, have a meal and talk about what's been happening. There is a social side to the group's success, because the Esperance community is pretty isolated and has developed an independent spirit that serves the group well. They meet a few times a year, talk things over, make a decision—and vote on it, if necessary.

Having a good committee is important because it gives the Group Leader support—he can make the noise up the front, but then the others follow through. It's been helpful to have DPAW and DAFWA representatives at the committee meetings—they lay the law down when it's needed. Over time, a good relationship has developed between the group and the government representatives, there is a feeling of mutual respect. It's professionalism that counts, being able to put the business of the group above self-interest. People may want to get stuck into government staff about some issues but a good chairperson can instruct all members of the meeting to not get personal—that's the level of professionalism that a good leader can carry.

The leader's facilitation of the group is part of that, and also his "no dickhead policy" which is, if you're a troublesome or cumbersome person who interferes or isn't on the same path and contradicts and tries to derail direction or progress, then you're not going be part of that group. That's how they make great decisions, because everyone thinks similarly and it's just the finer details that have to be worked out through good conversation. Everyone on the committee is go-ahead: they don't wait for government to tell them what to do.

Some people are scared of getting involved in committees—they feel like they have to take on responsibilities, or be there for the next 10 years. They don't see that the more people you have in those groups, the more options you have. The group does need to think about succession planning, to build up the confidence of the other members to do the job. There are intelligent men and women there, and they make good decisions as a group. Strong leadership is the key ingredient in the group's success so far. The leader has the passion and takes control, because it affects his livelihood but if he steps down, it's not clear who would be able to step into his shoes.

It's a well-structured group but the members still hold the view that they're doing the government's job for them. Landholders often declare that the dogs are coming from the public lands and the group is doing the work on behalf of the government. But it is the legal responsibility of all landholders to control wild dogs. It's a change from the old Agriculture Protection Board days, when DAFWA had a large number of staff and the government coordinated pest control. Now DAFWA will only work with the group in a formal partnership. The group knows that the change is not a joke, and there has been recognition that DAFWA is serious about it.

While the group agrees that dogs have a heavy impact on sheep production and that emus can cause damage to crops, government maintains that accusations that all feral animals are coming out of the public lands are not right. Feral animals roam everywhere: they're within the road reserves, in the corridors, in the vegetated strips along the river, within the national parks—they're on farmland, they're really everywhere. So while the group is quite happy to build a fence on the basis that the wild dogs, emus and dingoes can stay on "the other side" of the fence and farmers can do what they want on "their" side, this is not widely accepted as the only approach to the issue. For example, the Dingo Conservation Society don't support dog control measures and sponsored a trial of Maremma dogs, which was not regarded as a success by the NMDSG group leadership. Scott then asked them to contribute to the costs of building the fence, but there was no response to that request.

Although everyone blames the decline of the sheep industry on wild dogs, it is possible that the wool price is part of the problem. Dogs have had a big impact, but also sheep are more labour intensive than crops, and the younger folks seem to like cropping more. They can sit in their tractors and get on their phones and organise their marketing and that sort of stuff while they're working. The NMDSG knows that if sheep farming is to survive, they need to keep the pressure up. There is a definite disconnect between the city and country—city people would never contemplate what pest control is like, going out there day after day and putting hundreds of baits down just to kill animals. But this is what has to be done if the public want to keep lamb prices down.

The long-term strategy is to get the fence up—it is the only solution as far as Scott Pickering is concerned, and both he and the Treasurer are in it to the end.

20.3 Narrative Analysis—Northern Mallee Declared Species Group (NMDSG)

This story takes place in the context of substantial government reform. Redirected government budgets have led to the disruption of pre-existing relationships between operational staff and community members. These disruptions challenge coordinated action across public and private lands, and make it hard to build trusting relationships. Less government funding to both DAFWA and DPAW has also reduced wild dog control and other land management activities on public lands. Sheep farmers in the district report increased vulnerability to wild dog impacts as a result and feel they are bearing an unreasonable responsibility for managing this threat. Community action on wild dogs around Esperance had previously received financial support from the state government DAFWA and DPAW, with NMDSG accepting coordination and reporting responsibilities for a trapping and baiting program. Due to the implementation of the *Biosecurity and Agriculture Management Act 2007* (WA), this model has changed and the NMDSG is experiencing pressure to reorient from a single species focus to a wider biosecurity group.[3]

This change flags increasing administrative duties for the group, including writing funding applications and associated accountability requirements. Limited information about these changes and slowly increasing government expectations have created a power imbalance that burdens committee members while simultaneously promoting a message of community leadership and ownership. The group has already gained experience in the formal aspects of community-led action including incorporating as an association. As a result of the well-coordinated and sustained wild dog control activities organised by the group, they have gained legitimacy and the trust of the local community. In combination, these elements suggest that the group is well placed to meet the challenges of the new model. However, it is also possible that the group's single-minded focus on wild dog management may not easily transfer into a broader biosecurity context.

The NMDSG members see sheep farming as a way of life that is under threat. Because sheep farming is labour intensive, farmers' investment in their flock is not just financial. Wild dog attacks have a significant emotional impact on the farmers and when this combines with a loss of production income from wild dog attacks, falling wool prices or declining environmental conditions, they are faced with a choice to take action or walk away from the industry. The effort of establishing bloodline and breeding is a significant issue in some cases as years and years of work are potentially down the drain. An emotional undercurrent of inequity and unfair treatment for the sheep farming community of the Esperance region runs through this story. The farmers see themselves as independent pioneers responsible for developing the economic potential of the area. More established agricultural

---

[3]The Northern Mallee Declared Species Group was recognised by the Minister for Agriculture and Food (Western Australia) on 23 February 2017, and was incorporated as the Esperance Biosecurity Association (Inc.) on 18 May 2017.

areas are seen to have benefitted over a longer time from government investment and services that are now being wound back.

In this context, the group members see themselves as picking up the slack on behalf of the wider public. These increasing responsibilities are seen as evidence of a government retreat from pest control in WA. Farmers' historic reliance on DAFWA staff to plan and implement invasive species control has been replaced by an emphasis on landholders' legal responsibility to control declared pests such as wild dogs and government reform of the community funding model.[4] The line between abandonment by government and community independence is blurred in this story.

The story of the NMDSG hinges on the personal drive of the main protagonist, group instigator and leader, Scott Pickering. Scott is well connected in the sheep industry, forceful in his views and confident in his leadership role. He has recruited a committee of allies who support his vision and this group has been recognised as successful because of their internal cohesion, as well as their ability to organise and sustain action on wild dogs. The leader embraces the political dimensions of the issue and actively participates in industry and government mechanisms to influence decision-making. Under this politically astute leadership, the NMDSG has developed a campaign focused on extending the state barrier fence to prevent wild dogs moving into the agricultural areas of the region. The ability of the group to get the fence proposal taken seriously adds to their reputation as a successful group.[5] Government stakeholders point to the group's tenacity, endurance and focus as strengthening their recognition and standing in the wild dog management context. A good example is the group's successful reframing of the wild dog problem to include emu impacts on cropping areas, as a strategy to bolster cross-industry support for the barrier fence.

Common goals and strong leadership have been key ingredients for success in this case. The leader has gained the respect of all stakeholders interviewed for this research. However, over-reliance on a strong leader can be a risky strategy for a community group, particularly when circumstances require leadership change. The group does not seem to have a strategy in place for generational change, and this creates a challenge to the long-term sustainability of the group. Successful completion of the fence extension project has become the timeframe for any serious consideration of group leadership change.

Group members value the strong sense of belonging and unified purpose that the group leadership team actively cultivates. Routines that combine meetings with social activities create strong personal bonds between group members. As a result, they rarely disagree and decisions are usually unanimous. This cohesion is not accidental; it is actively developed and defended, with dissenting voices challenged or dismissed from the group. As a result, the group membership represents a particular version of community that may reduce the pool of possible candidates for committee roles.

---

[4]Recognised Biosecurity Groups can raise a pest management rate which is eligible for matched state government funding under the conditions of the Biosecurity and Agriculture Management Act.

[5]In an announcement on 9 November 2017, the WA government committed 60% of construction costs for the barrier fence extension project. It is unclear if the remainder of the project funds will be secured.

This is a challenge for an incorporated group, particularly as not everyone has the necessary skills or feels capable of being actively involved in group membership.

The NMDSG presents a version of successful community-led action that relies heavily on political influence and strong leadership qualities. There are many positive lessons from this case: unity of purpose, group cohesion and active leaders are ingredients for successful action. However, there are alternative views of what is happening in the landscape and the group's capacity to accommodate these is questionable. The devotion to the fence extension may result in alienation of dissenting viewpoints. It may also miss the possibility of using biodiversity values to motivate community-led collective action for wild dog control.

Throughout this case, a perceived imbalance between government control of resources and the group's ability to unlock these heighten tensions. The quality of the community-led action is complicated by this tension, which suggests that there is a dysfunctional power dynamic being reinforced by the top-down reform of biosecurity management. There is a clear advantage to the government in driving a community-led model—they can divest responsibility, reduce funds and create an imperative for action. However, the advantages for the community group are less compelling, as increasing administration costs and accountability requirements on volunteer time create confusion about where the real benefits of "community-led action" lie in this case.

In response to this perceived power imbalance, the group leaders are prone to issuing ultimatums that threaten a stop to community-led action on wild dogs, despite acknowledging that this would most likely harm sheep producers more than government representatives. It is unclear what impact the adversarial tone of the group leadership has on broader community engagement in the wild dog program. This does not take away from the strong emotional and financial motivation that drives group members to participate in collective action. Ultimately, they want to prevent others from being subject to the trauma of wild dog attacks on their sheep. It is possible that the increasing workload associated with regular and maintained dog control makes the prospect of a barrier fence seem like a permanent solution that will solve more than just the dog control problems.

References

Australian Bureau of Statistics. (2011). *Cat. No. 7121.0—Agricultural commodities, Australia, 2010–11*. Canberra, ACT: Australian Bureau of Statistics.

Department of Agriculture and Food. (n.d.). *State barrier fence extension update*. https://www.agric.wa.gov.au/invasive-species/state-barrier-fence-esperance-extension-update. Accessed February 22, 2016.

Department of Agriculture and Food. (n.d.). *State barrier fence overview*. https://www.agric.wa.gov.au/invasive-species/state-barrier-fence-overview. Accessed February 22, 2016.

Department of Agriculture and Food. (n.d). *State barrier fence R4R projects*. https://www.agric.wa.gov.au/r4r/state-barrier-fence-r4r-projects. Accessed February 22, 2016.

Department of Agriculture and Food. (n.d.). *Western Australian organism list*. https://www.agric.wa.gov.au/organisms. Accessed February 22, 2016.

Department of Agriculture and Food. (2014). *Wild dogs*. https://www.agric.wa.gov.au/state-barrier-fence/wild-dogs. Accessed February 22, 2016.

Department of Agriculture and Food. (2015). *Recognised Biosecurity Groups (RBGs)*. https://www.agric.wa.gov.au/bam/recognised-biosecurity-groups-rbgs. Accessed February 22, 2016.

Department of Parks and Wildlife. (n.d.). *Pests and diseases*. https://www.dpaw.wa.gov.au/management/pests-diseases. Accessed February 22, 2016.

Driscoll, D. (2013, February 27). All cost, little benefit: WA's barrier fence is bad news for biodiversity. *The Conversation*. https://theconversation.com/all-cost-little-benefit-was-barrier-fence-is-bad-news-for-biodiversity-12333. Accessed February 22, 2016.

Johnston, B. (2011, January 1 date). Maremma dogs provide wild dog solution in WA. *Stock & Land*. http://www.stockandland.com.au/story/3561907/maremma-dogs-provide-wild-dog-solution-in-wa/. Accessed February 22, 2016.

Jose, L. (2014, July 4). Western Australian declared pest control funding to change. *ABC Rural, WA Country Hour*. Australian Broadcasting Corporation. http://www.abc.net.au/news/2014-07-04/declared-species-money/5572530. Accessed February 22, 2016.

Local Government Network, Australia-LGNet. (n.d.). *Shire of Esperance*. http://www.esperance.wa.gov.au. Accessed February 19, 2016.

Natural Resource Management Program. (n.d.). *11050: Wild dog management to protect the Esperance farming region*. http://www.nrm.wa.gov.au/projects/11050.aspx. Accessed February 22, 2016.

Northern Mallee Declared Species Group Inc. (2011). *State barrier fence—Esperance extension, farmers' meetings presentation*. Resource document. http://northernmalleedsg.org.au. Accessed February 22, 2016

Northern Mallee Declared Species Group Inc. (2014a). *Become a member*. http://northernmalleedsg.org.au/become-a-member/. Accessed February 22, 2016.

Northern Mallee Declared Species Group Inc. (2014b). *Contacts*. http://northernmalleedsg.org.au/contacts/. Accessed February 22, 2016.

Northern Mallee Declared Species Group Inc. (2014c). *Frequently asked questions*. http://northernmalleedsg.org.au/faqs/. Accessed February 22, 2016.

Northern Mallee Declared Species Group Inc. (2014d). *History*. http://northernmalleedsg.org.au/history/. Accessed February 22, 2016.

Shire of Esperance. (2015). *Ordinary Council Agenda, 27 October: Request for assistance from the Northern Mallee Declared Species Group*. Council document. http://www.esperance.wa.gov.au/Infocouncil/Open/2015/10/ORD_27102015_MIN.HTM. Accessed February 22, 2016.

South Coast Natural Resource Management. (n.d.). *Wild dog control in the Shire of Esperance*. http://southcoastnrm.com.au/item/wild-dog-control-in-the-shire-of-esperance. Accessed February 22, 2016.

Thompson, P. (2008). *Farmnote: Wild dog control*. Forrestfield, WA: Department of Agriculture.

Visit Esperance. (n.d.). History. http://visitesperance.com/pages/history/. Accessed February 19, 2016.

WA Country Hour Team. (2012, April 16). Wild dog numbers in WA force some farmers out of livestock and PGA calls for bounty. *ABC Rural, WA Country Hour*. Australian Broadcasting Corporation. http://www.abc.net.au/site-archive/rural/content/2012/s3478621.htm. Accessed February 22, 2016.

Wikipedia. (2016). *Esperance, Western Australia*. https://en.wikipedia.org/wiki/Esperance,_Western_Australia. Accessed February 19, 2016.

Chapter 21
Three Wild Dog Group Case Studies: A Meta-analysis

Abstract In this chapter, we investigate some of the common themes that emerge in the wild dog group case studies. These include:

- The emotional dimensions of wild dog management—how positive and negative emotions influence a community response to wild dog threats.
- The capacity of an affected community to act—how different models of decision-making, levels of support from government or industry, skills and financial resources in the affected community, and the ability to influence policymakers, shape the community response.
- The importance of leadership and community structure—in particular, the role of a willing leader and supportive members; the creation of shared experiences and a common purpose; and characteristics of determination and persistence.
- The role of power and influence—the successful community groups were able to access important information and share it, breaking down power imbalances, and empowering their members.
- Naming and framing the issue—developing a shared understanding of the problem was essential to creating a shared vision for action; groups were more successful when they 'owned' the problem of wild dog management and co-created the solutions.

Before leaving the case studies, we briefly consider all three together to deduce what lessons might be drawn from these stories, and how they might inform the development, planning and implementation of community pest management initiatives. This analysis and discussion are anchored in our conceptual framing of community pest management as a problem that requires the development of a *shared vision* of both the threat and possible solutions, so community stakeholders can then make a *shared commitment* to take collective action for pest management. In this way, the policy principle of shared responsibility becomes more meaningful and attainable for

T. M. Howard et al., *Community Pest Management in Practice*,
https://doi.org/10.1007/978-981-13-2742-1_21

all stakeholders involved in community pest management. Many stakeholders likely don't always think in the policy language or principles of "shared responsibility"; yet when people work together across differences and take ownership for their role in pest management, they are actively shaping a shared response to pest threats. In the following paragraphs, we consider some of the cross-cutting social dynamics that the narrative enquiry revealed about community pest management.

21.1 Emotional Dimensions

There is no doubt that emotion, particularly for sheep farmers, is a significant feature of these wild dog management narratives. Dogs have a real personality in each case. Regarded as smart, devious and wilfully destructive, they seem to delight in the heartbreak they cause the sheep farmer. The farmers and wild dog controllers regard the dogs with grudging respect, and see them as a genuine threat to their way of life. This creates a highly charged context for wild dog management which had implications for attempts to achieve collective community-led action. A community that feels under threat may react in different ways, expressing fear, passion and anger in relation to wild dog management. Sometimes, these feelings are focused on the dog itself, but they are also likely to spill over into social or organisational settings. Based on these emotions, individuals may develop positive or negative relationships with other community members, government employees or industry representatives. Several farmers across the cases expressed their situation as one of conflict: with the dogs, the government, and in some cases, with their neighbours and other community members.

Where barriers or impediments such as baiting regulations were experienced, frustration fuelled increased antagonism to authority figures. The common complaint that government did not understand the needs of the farmer reflected a feeling of isolation that stoked fears of abandonment. Farmers in these cases faced a threat to their livelihood and the stakes were correspondingly high. The realisation that there is no permanent solution to the "problem" of wild dogs had prompted some to walk away from sheep farming altogether. Seeing others leave the industry raised the stakes for the remaining farmers, who either sought support from others or withdrew from community action in an expression of defiance.

Change fuelled emotions, prompting innovation or entrenchment.

In each case, farmers expressed concern about the changing circumstances they found themselves in. They were experiencing a change from one way of life to another, with a notable increase in absentee landholders, amalgamation of small farms into bigger properties, and in the peri-urban context of Mount Mee, a decreasing use of land for agricultural production. Most farmers had long family connections to the region or the sheep industry, and could remember the heyday "when wool was king" (Victorian narrative). The industry now seemed precarious and increasingly vulnerable to unsympathetic policy and increased dog impacts. Concern about the future of their own enterprise was usually linked to worries about the economic and social survival of the surrounding rural community. Each case also hinted at an existential fear of rural invisibility—the sense that decisions about rural livelihoods were being made by urban voters or policymakers with limited understanding of the conditions faced by those living in country areas. There was a noticeable tension between fear of change and the need to innovate, which can be difficult to balance in conservative rural communities where new ideas are challenged by tradition or entrenched behaviour. Old ways of thinking about wild dog management had been passed down through the generations and this was particularly influential in Swifts Creek, where limited social mobility and a lack of outside experience created a version of local knowledge that struggled to adapt to change.

Trust affected actions.

Another common emotional thread was the importance of trust in shaping the way the affected community met the wild dog threat. Community leaders that had the full support of their community were regarded as trustworthy. Government employees were often regarded with distrust, even when the individual was not directly responsible for regulations or past experiences. Building trust was linked to respectful interpersonal interactions that gave community members the opportunity to be heard and acknowledged. It was also linked to evidence of supportive action, such as providing coordination for baiting programs, or minimising paperwork to meet regulations. Feelings of distrust and antagonism were particularly strong where the government was seen to be applying top-down pressure on community groups to take responsibility for wild dog management. In this context, non-government coordinators such as industry facilitators were more favourably regarded. Sometimes, this dynamic intensified conflict between stakeholders, leaving individual practitioners feeling exposed to community criticism. These tensions could lead to perverse consequences, as government or industry staff attempted to prove their trustworthiness by expressing personal dissatisfaction and cynicism about the policy framework, thereby identifying themselves as "outsiders" to the government process and an "insider" to the community itself. This dissonance at the front line is concerning because *"trust is the core link between social capital and collective action"* (Ostrom and Ahn 2007, p. 8) and without trusting relationships between all affected stakeholders, achieving the principle of shared responsibility becomes less likely.

21.2   Capacity to Act

Capacity to act emerged as a significant factor in each case, and was crucial to the potential for collective community action. Each group had different needs which were directly linked to the capabilities of the landholder community. This reminds us that while on one level, wild dog control is about taking on-ground action, on another level it is also about the capacity of the community to participate in either formal or informal models of collective action. The cases illustrated diverse models of decision-making and community organisation. Only the Northern Mallee Declared Species Group (NMDSG) had formally incorporated, a process that then raised volunteer concerns about the amount of time required to meet the required accountability requirements. Incorporation also required suitable candidates with the required skills to fill key committee positions, which was a challenge in sparsely populated rural areas. The other cases were proudly informal, meeting as needed and relying on the established networks of the "bush telegraph" to stay in touch.

External support was often beneficial in increasing capacity.

Regardless of their status, each case benefitted from significant levels of government, non-government and industry support. Coordination of baiting, assistance with meetings, financial reporting and information services provided by external stakeholders were present in each case, to varying degrees. The extent to which these support functions were recognised and acknowledged by the community group leaders and members depended on the quality of personal relationships and management history in each case. It is possible that the farmers who expressed pride in their independent, self-reliant image may have underestimated the contributions made by other actors in support of community action on wild dogs. The case studies did reveal a range of policy and program design features that supported community members to navigate bureaucratic procedures and build confidence in dealing with policy. Examples included providing feedback on funding applications in NMDSG; reducing regulatory barriers through streamlined baiting systems at Mount Mee; and building confidence at Swifts Creek through relationship development and information provision. In this way, external stakeholders added value to the wild dog management system by bolstering landholder skills and capacity.

Communities faced sustainability questions in balancing external support with local capacity.

The Mount Mee example raised an interesting question about the interplay between community interests and external support. At Mount Mee, stakeholders identified coordination services supplied by the local government as key to the success of the management program. Thinking critically about this model of government-community partnership, we ask whether the community action achieved would be sustainable if the coordination service ceased due to change of policy or budgets? The recent history of this particular case suggests that the challenges of coordination would once again overwhelm the local community's capacity to act effectively, leading to a fragmentation of effort that would soon see the wild dog problem re-emerge. In this specific example, we observe a challenge that applies equally to the other cases and to other problems that require sustained collective community action. Finding the balance between external service provision and support, and community-generated effort and action, is a persistent challenge for community pest management.

21.3 Leadership and Community Structure

Each case identified a community leader that had been influential in getting the group started. Although leadership style and influence varied, it always involved significant investment of personal time and effort. The leader seldom acted alone but was supported by a team of other landholders who provided both practical support and moral support. They amplified and reinforced the leader's activity by getting involved in the control program and maintaining their involvement over time. In both NMDSG and Ensay, the leadership team included farmers who had moved out of sheep farming but saw their involvement as a community contribution, a civic duty that reinforced good neighbour behaviours in the rest of the group. The ability of leaders to negotiate mutually beneficial outcomes seemed largely dependent on their skills and their access to resources. Where leaders had good political networks or industry influence, there was a greater engagement with policy and negotiation of responsibilities. This was most clear in the Victorian case, where neighbouring communities responded to similar challenges in very different ways.

Participatory leadership strengthened community-led action.

Good leadership also had the potential to create an alignment between community values, management objectives and on-ground activities. At Mount Mee, alignment allowed a self-reinforcing wild dog management system to emerge over time. In this case, processes such as collaborative planning and nil-tenure mapping involved leaders from each stakeholder group in developing a management plan. Through that process, individuals then identified and resolved conflict of objectives, possible problems with coordination and resourcing, and increased the possibility of shared ownership. This participatory process required strong leadership that could identify coalitions and build interest networks that aligned with the needs and interests of the affected community. The importance of community leadership in these cases suggests that leadership development could be an important strategy for strengthening community-led action for pest management, particularly in diverse socio-economic contexts.

Strong leadership sometimes had adverse consequences and raises questions for communities.

While the cases showed that strong leadership was a necessary ingredient for inspiring community action, there was evidence that it could also have unintended side effects. For example, while forceful leadership was identified as an asset for the NMDSG, it also created tensions when dissenting voices or perceived outsiders tried to participate. It is not clear if there were negative consequences for the NMDSG from the Chairperson's lack of tolerance for those who he deemed as unhelpful or antagonistic to the group's aim. The effectiveness of the NMDSG group was also linked to internal social cohesion which was actively developed by the leadership team. A possible unintended consequence was the lack of a succession pathway due to the strong reliance on the leader's personal drive and commitment. This created an ongoing risk for the program, as it increased vulnerability to sudden and abrupt leadership change. Although this risk was recognised by the committee, there was no active strategy in place to avoid this possible outcome.

Shared experiences were necessary in growing a community of leaders.

Beyond strong leadership, community action in wild dog management requires a platform of shared experience and understanding in order to develop shared commitment. Although the cases presented here illustrate different versions of community pest management, they also shared several significant features. The landholders in these case studies were largely older (over 50 years) Caucasian men. The economic viability of their surrounding rural community was a motivating concern for all parties. They saw their rural communities as vulnerable to events which impacted negatively on agricultural production, such as wild dogs. Landholders often shared historical ties and physical boundaries. In some instances, these close social and geographic connections facilitated good working relationships and lowered conflict between landholders. The dark side of this coin was seen in the Victorian case, where historical enmity between neighbours resulted in questions about the legitimacy of some landholders to make decisions about dog control actions. A similar challenge was reported in the peri-urban context of Mount Mee, where a growing number of landholders were no longer involved in agriculture. In this case, participation in coordinated baiting efforts was built through active community leadership to reach out to these non-agricultural interests and raise awareness of the shared impact of the issue.

21.4 Power and Influence

Power and influence took different forms in each case; each form is important to consider to better understand how power shapes community responses to pests. To understand power in action, it is necessary to examine how the benefits and costs of wild dog control are distributed in any particular community; and to critically assess program and policy design to ensure that a "one-size-fits all" approach to community development does not reinforce existing patterns of socio-economic advantage. Bureaucratic power appeared as both a perceived and genuine threat to community-led action, with regulations, funding flows and changing policy objectives creating instability and uncertainty for community members. At the same time, each case demonstrated a power struggle between the individual landholders' desire for independence, and a tendency to look to government for solutions. In all cases, individuals spoke of power tussles that were influential in the way they perceived the problem and how they conceived of solutions. This was most obvious in Swifts Creek, where socio-economic disadvantage fuelled a strong sense of injustice linked to dynamics of exclusion from power. Those community members felt their concerns were viewed as "whinging", rather than as raising legitimate questions about the influence of privilege. In contrast, political power was successfully employed by Ensay and NMDSG to take control of some parts of the agenda. Achieving changes to baiting restrictions in the Ensay case was empowering and motivating for the group members; the ability of the NMDSG to get the fence-extension proposal taken seriously demonstrated the group's political influence. Although both examples increased empowerment for the landholder group, the extent to which they created new models of power sharing that might be sustained over time remains unclear.

Perceptions of different knowledge sources affected actions taken and how people worked together.

Controlling access to knowledge is another manifestation of power. If knowledge is held by only a few, they have a disproportionate ability to set the agenda. Selecting which piece of knowledge to share through media, meeting documents, or interpersonal interactions, facilitates a power imbalance. The case studies demonstrated both positive and negative implications of this phenomenon: successful leaders were seen to have access to information, skills and influential networks that may have been withheld from others in the community;

This link between knowledge and power was clearly identified in the Victorian case studies. Farmers felt that local knowledge could make a valuable contribution to wild dog management, however, legislation and policy knowledge was concentrated in selected public servants or landholders, creating a sense of disempowerment and exclusion for others. A subset of Swifts Creek landholders identified that access to good quality information was the key to achieving systemic change. Through local knowledge and community-driven innovation, they developed an improved data system that could provide landholders with the information they needed to reform the system, improve their practice and get funding. As a result of this community effort, landholders drove a culture of increased professionalism in the government wild dog program. Participants are now involved in data collection, recording GPS coordinates of their bait runs and completing accreditation to lay baits themselves. These are all examples of knowledge sharing which leads to landholder empowerment.

Sharing knowledge is a key strategy for breaking down power imbalances. Mechanisms such as nil-tenure planning forums, advisory groups and local community visits can increase opportunities for power sharing among stakeholders. This was demonstrated at Mount Mee, where the use of maps connected landholders in the landscape and made it easier to think about how the dogs were moving across the whole area. This facilitated sharing of local knowledge was valuable for both understanding the landscape, and learning more about the motivating drivers for community participation. Regular visits to individual farms and use of video data were ways that Mount Mee Council staff could share their knowledge in a non-threatening situation, and invite landholders to contribute their local expertise.

By examining these stories of collective action for wild dog management, it is possible to uncover forms of community power in rural areas that are seldom recognised or described. Landholders that have successfully organised to take control of their wild dog problem are demonstrating how "people power" can transform a complex problem into a collective effort. Participatory planning processes, such as the nil-tenure approach, offer pathways to change established power relationships. This requires neutral parties to broker relationships across historic divides and address underlying power struggles. These cases show that while community power may be difficult to recognise and incorporate into the decision-making processes of a public policy framework, it does exist and has real impacts on ground.

21.5 "Naming and Framing" Issues in Wild Dog Management

Across the case studies, the link between individual value systems, ways of knowing and access to information influenced the way wild dog management was perceived and described. This included the way that public and private benefits of wild dog management were "named and framed". Farmers often expressed a libertarian desire to be free of government interference and regulation; at the same time, they called for increased government management of the wild dog problem. Public lands were seen as a problematic source of wild dogs, with the impacts disproportionately felt by sheep farmers. While public land managers recognised these concerns, they saw the issue as one of biodiversity impacts rather than agricultural livelihoods. This revealed that the "problem" of wild dog management in these cases was predominantly named and framed in relation to livestock (and particularly sheep). The livestock industry had an economic interest in reducing wild dog impacts, and this influenced their definition of the problem and the possible solutions.

Throughout the case studies, various visions of "success" and "failure" for community action for wild dog management emerged. Some definitions of success were linked to:

- Reduced wild dog impacts on stock
- Greater social cohesion and community building
- Community acceptance
- Widespread support
- Funding
- Political recognition and
- Enjoyable social interaction.

Communities showed that evaluating success is complex.

In Ensay, increasing awareness of animal husbandry and a change of focus from dog control to herd health saw sheep numbers increase. This range of measures suggests that ideas of successful community action are linked to individual perspectives of what the benefits are, and where they accrue. Definitions of success are informed by complex interactions between the social, the emotional and the political. The way an issue is described and *by whom* will influence the version of success that is used as a benchmark. This matter is also a function of power, our last theme. Naming and framing of the problem and what constitutes success are illustrative of the power dynamics in any given context. These functions may be disguised within mundane or bureaucratic processes of project definition and administration. Power imbalances lead to problematic interdependencies, within the community groups themselves, and in relations between external and internal stakeholders. This is important because whoever gets to define the metrics of success or effectiveness also has the power to frame the nature of the problem or issue of focus. Like many issues related to power, these imbalances can be hard to spot unless the voices of those affected by the imbalance are actively sought out, listened to and *heard*.

The case studies revealed that it was too simplistic to rate a collective action initiative as either a "success" or "failure". An important indicator of success in wild dog management is the sustainability of the community-led effort over time. The continuing presence of the threat with no possibility of eradication creates an ongoing requirement for community action to keep the dog population under control. The long history of community leadership in the Mount Mee example, and the ability to learn and adapt, suggest that resilience and adaptability were key measures of success in this case. In Ensay and NMDSG, high levels of interpersonal trust and peer recognition were crucial for building group resilience to barriers, and encouraging sustained participation. For Swifts Creek, the sustained effort was seen as a battle for justice and recognition, although this led to a battle-weary community.

Frames included peer recognition as a form of success.

Peer recognition was an important indicator of success in all cases. This took the form of acknowledgement by neighbours or other community members. Recognition by local media helped build the profile of the collective effort, and media was often used by group leaders to prod government into recognising the seriousness of their experiences with wild dog impacts. For Ensay and NMDSG leaders, nomination to various advisory bodies as a landholder or industry representative was identified as a form of political recognition. These forms of recognition increased the individual's reputation and by association, raised the profile of the collective effort. This is known as "reputational politics" (Fine 2012) a valuable asset for community groups. Recognition strategies may be important for maintaining motivation to both take responsibility for the pest, and participate in a collective effort.

Frames can differ from stakeholder to stakeholder and can change collectively within communities.

"Doing the government's job for them" (Swifts Creek narrative) was a common negative framing of landholder efforts in NMDSG and Swifts Creek. However, this transfer of responsibility was reframed as a positive by the Ensay group, who saw an opportunity to take control of the baiting program and drive the agenda according to their needs. The *problem* then became an *opportunity*, a source of group empowerment which reinforced feelings of success. Ensay also reframed wild dog management as one of many factors affecting production, and integrated a range of information services under the general banner of better sheep production. The early stalled effort at Mount Mee showed how challenging wild dog control programs can be, particularly when individuals feel unsure about how the control will impact on their livestock, domestic animals or personal health. Baiting needed to be reframed as a safe and reliable method of dog control, well regulated and effective, in order to build community trust and support for the program. This reminds us of the need to look beneath the surface of how an issue is described to consider who is telling the story and what perspective they bring. The wild dog case study narratives remind us that achieving community action requires care and consideration of different values in order to build trust and encourage sustained participation.

21.6 Conclusion

Wild dogs exist in a complex biophysical and pyscho-social landscape (Fitzgerald et al. 2007). The case studies demonstrate that the recognition and acceptance of a wild dog threat are not sufficient to create a collective community response. There is a complex social dynamic of community formation that underpins progress from threat awareness to collective action (Everts 2015) and this must be navigated with care, acknowledging the emotional aspects of perceived problems and proposed solutions. We've noted several of these complexities throughout the chapter:

- Emotions impact actions, yet those emotions don't operate the same for every individual or community.
- Pest issues can mean change for many individuals, which can fuel emotions leading to innovation or entrenchment.
- Trust affects actions.
- External support showed beneficial in increasing community capacity, but raises questions of sustainability.
- Participator leadership strengthened community-led action, while strong centralised leadership sometimes brought adverse consequences on shared capacity.
- Shared experiences were necessary in these cases.
- Knowledge is a form of power; which knowledge sources are recognised and valued can impact what actions people take and how people act with one another.
- Definitions of success vary and evaluating success is complex, yet all frames seen in the cases show peer recognition as one indicator of success.
- Recognising different frames of one issue can increase understanding of collective issues and ways to coordinate shared visions and actions.

The case narratives identify and illuminate power dynamics, hierarchies and asymmetries that are important for understanding questions of community pest management. These dynamics can either support cooperation between diverse parties or lead to conflict which reduces the potential for collaborative effort. Where conflict arises, it intensifies the powerful emotions associated with wild dog impacts and leads to feelings of helplessness that may manifest as antagonism to government, other communities or even neighbours—particularly when these other parties are seen to possess more power. In all three cases, landholders had a desire to be heard, influence decisions and drive their own destiny. These are necessary motivators for community action and deserve increased attention in pest management programs.

References

Everts, J. (2015). Invasive Life, communities of practice, and communities of fate. *Geografiska Annaler: Series B, Human Geography, 97*(2), 195–208.

Fine, G. A. (2012). *Tiny publics: A theory of group action and culture.* New York, NY: Russell Sage Foundation.

Fitzgerald, G., Fitzgerald, N., & Davidson, C. (2007). *Public attitudes towards invasive animals and their impacts.* Literature Review. Invasive Animals Cooperative Research Centre. https://www.pestsmart.org.au/public-attitudes-towards-invasive-animals-and-their-impacts/. Accessed December 12, 2017.

Ostrom, E., & Ahn, T. K. (2007). The meaning of social capital and its link to collective action. In G. T. Svendsen & G. L. Svendsen (Eds.), *Handbook on social capital* (pp. 1–34). Northampton, MA: Edward Elgar.

References

Part III
Learning from Stories of Practice

Chapter 22
Conclusions

Abstract In the opening chapters of this book, we observed that the issue of pest management is not only about species ecology and management technologies. It is a complex, persistent and sometimes wicked public issue that requires human action. In this final chapter we revisit the role that narrative plays in understanding the human dimensions that are influential in shaping collective responses to invasive species management.

In Chap. 2, we spent some time outlining the theory of narrative enquiry that underpins this book. In this final chapter, we reaffirm our earlier agreement with Mayer (2014) who wrote that:

> Narrative is perhaps *the* essential human tool for collective action, a tool of enormous power and flexibility for constructing shared purposes, making participation in collective action an affirmation of personal identity, providing assurance that others will join us in the cause, and choreographing coordinated acts of meaning. (Mayer 2014, p. 49)

This book has been framed around the power of stories and their utility for understanding and addressing thorny issues of collective community action such as pest management. The practitioner profiles and wild dog management group case studies contribute to long-running efforts to increase understanding of the interaction between the social and scientific dimensions of pest management. For example, back in 1994 the Nature Conservation Council of NSW convened a conference on the topic of *Unwanted Aliens: Australia's Introduced Animals*. The conference was framed around an acknowledgement that "control techniques [are] not applied in a social and political vacuum" (Diekman 1994), and along with the scientific research that identified the serious impact of a range of invasive species, several community leaders and landholders spoke about their experience of introduced animals. Roland Breckwoldt, a landholder from Bega, NSW, pointed out that "all of us who eat grain or beef or both, wear wool or cotton or both, are part of the same system that has within it the introduced animal pests" (as cited in Diekman 1994, p. 97). After positioning humans thus within the pest management system, Breckwoldt went on to emphasise the role of stories and myths in shaping human attitudes to particular species, noting that "such ideas are powerful and do gain a foothold in the public perception",

© Springer Nature Singapore Pte Ltd. 2019
T. M. Howard et al., *Community Pest Management in Practice*,
https://doi.org/10.1007/978-981-13-2742-1_22

impacting on management regimes and influencing community responses (as cited in Diekman 1994, p. 96).

It will not surprise our reader to know that we agree with Breckwoldt's suggestion that myths and stories have an important role to play in the design and implementation of community pest management efforts. Our interest in narrative has emerged from the belief that human understanding is constructed and mediated through the telling and re-telling of stories. However, this is not our only interest. As citizens and scholars concerned with reinvigorating and strengthening the theory and practice of community development and engagement, we also want to make a strong statement about the value of this approach and the contribution it can make to community pest management.

In the opening chapters of this book, we observed that the issue of pest management is not only about species ecology and management technologies. It is a complex, persistent and sometimes wicked public issue that requires human action. Where pests have become widespread and established in the landscape, the efforts of individual citizens are not enough and effective pest management requires a collective effort. Our interest has been to understand the social and political process of community formation and action through the lived experience and first-hand accounts of individuals involved in community pest management. Through these narrative accounts, we have seen that collective action requires trust, respectful interpersonal relationships and reciprocity among those individuals who need to work together to achieve their individual and collective objectives. It seems to us that where these interactions reflect principles of democratic politics and practice, attempts to build shared purpose and trust amongst community members and other stakeholders are more effective and sustained over time.

Through our work with community engagement practitioners and affected communities, we have observed that although phrases like "community engagement" and "community-led action" are regularly used in pest management strategies, they often appear disconnected from serious thought about what such an aspiration might require in terms of human interaction, social and political change. These changes involve both personal commitments to sharing power and working across different viewpoints, and professional efforts to break down structural imbalances that devalue and marginalise community perspectives. This focus on human behaviour extends to all individuals involved in the collective effort, regardless of their organisational or political role. As a result, the significant change required to achieve genuine community management can be threatening to those who benefit from the status quo, and this may reduce the appetite for initiatives that explicitly aim to achieve this change.

For us, achieving community pest management is a political enterprise as much as a species management effort, because the nature of the problem requires that individuals work collaboratively to develop shared agendas, objectives and actions. This is not easy work.

There are many different human perspectives on the issue and these typically lead to conflict and resistance to pest management strategies. Personal values and a sense whether the role in decision making is genuine (or not), will influence community responses to scientific recommendations or management technologies. The influence of these social dynamics on the form and function of community pest management requires attention to both the biophysical nature of the pest and to individual and collective human behaviour.

The challenge is that both are necessary, but neither alone is sufficient.

Through our interactions with pest managers, land holders, community leaders and policy-makers we have also observed that many people are for some reason uncomfortable with framing community pest management as a political act, despite their personal and professional experience of political dynamics. This reluctance to think about community pest management in terms of politics, power, and democratic practice may be a result of the largely technical and scientific framing of the issue. In these brief conclusions, we will be explicitly addressing the interaction between politics and practice in order to develop what we see as a crucial platform for achieving community pest management—the awareness that it requires a shifting of power relations to enable leadership and action by both affected community members and those not directly affected, at both the individual and collective level.

In taking this stance, we are also making a political statement about the value of narrative research approaches in helping us understand the behavioural and social processes that underpin community formation and action. We are aware that this way of framing pest management is challenging, not just to the individual practitioner or community members, but also to the scientific paradigm that has dominated this field of research. However, as the many stories in this book illustrate, building community requires an awareness and acceptance of diverse values and ways of understanding in order to strengthen the collective capacity to act. There is no doubt that pest management remains a technical, ecological and biophysical challenge, however the increasing emphasis on principles of "shared responsibility" and a retreat from government intervention raises significant questions about the human dynamics of pest management. In other words, cooperation at the local level is increasingly important. As researchers, we see value in bringing attention to the tensions and imbalances in the current framing of community pest management, and strongly argue for a more genuine dialogue between the scientific and the social in both defining and addressing pest problems.

Looking back to 1994, we see that participants at the *Unwanted Aliens* conference were already interested in this interaction between the scientific and the social. They acknowledged that land managers (both public and private), practitioners and other stakeholders from government and industry hold diverse and nuanced perspectives about pest species management, stakeholder roles and responsibilities, collective action, and the barriers and opportunities for improvement in management practices.

Terry Korn, a vertebrate pest scientist, emphasised the language of "shared responsibility", a principle that still drives pest management policy today. We agree with his underpinning rationale that "strong ownership of a community decision generates positive action" (as cited in Diekman 1994, p. 108). Drawing on this analysis, we emphasise that to us,

> pest species management is essentially a human and community problem, and consequently, people and community, in the context of specific places or locales, must be at the centre of strategies to address this problem.

This agenda for change is at the heart of our book. As community members ourselves, we are inspired by the examples of social change already underway that are captured in these narratives. As researchers, we are motivated by the opportunities for renegotiation and restructuring of power dynamics and social structures that are evidenced by the practitioner profiles and community case studies. Our commitment to developing community engagement as a personal and professional practice has encouraged us to embrace active listening, spend time in critical reflection, strive for humility in our interpretation and above all, respect and value different perspectives. In the spirit of narrative enquiry, we once again invite our readers to assess the work on its merits, to critique and offer alternative interpretations—but above all, to engage with the serious political implications that these narratives reveal for furthering and sustaining the ambitious social and political change necessary to achieve community pest management in practice.

Everyday Politics

Before going further, we need to clarify what we mean when we talk about politics, as this informs our understanding of "political implications". The politics we refer to is not the formal mechanism of government, regulated through political parties and voting in a process of electoral representation. Rather, it is the daily lived experience of individual citizens engaged in routine decisions and practices of civil society, shaping their own lives and affecting those around them through their personal expression of values, beliefs and other guiding frameworks. As democracy theorist Harry Boyte writes, "[everyday] politics is the practice of power wielding to get things done in complex, heterogeneous societies" (Boyte 2004, p. 4).

It can be helpful to think about politics as either big "P" or small "p" versions of the human activity of decision-making and taking action (Boyte 2004). While only some of us are actively involved or interested in the machinations of big "P" politics such as elected government, we are all engaged in daily experiences of the small "p" version through our interactions with each other at both a personal and professional level. This is the everyday politics that each individual experiences through interaction with their community. Examples of this political experience are peppered throughout the practitioner profiles and the wild dog case studies. We see evidence of small "p" politics in the practitioners integration of community knowledge into the definition of the pest problem, and in the landholder suspicion of government bureaucrats in the wild dog case studies. Small "p" politics can influence efforts to build trusting and respectful interpersonal relationships because it lingers in community stories of past injustices or triumphs.

We previously observed that some people are uncomfortable with framing community pest management as an expression of politics. This reluctance may stem from a distaste for the power plays of big "P" politics and a sense of alienation from the mechanisms of political decision-making. This creates a tension between policy and program aspirations for community leadership, in pest management and other complex issues, and the lived experience of the individual citizen. For example, throughout the narratives we heard that individual citizens often feel overlooked or isolated in their efforts to manage pest species. "I didn't think anyone cared" was a common refrain from landholders, expressing an alienation from decision-making and power structures that undermined their participation in community pest management. This expression of alienation reminds us that the aspiration of shared responsibility must address the tension between small "p" and big "P" politics, in order to reposition the individual citizen as a civic leader and change agent within their own sphere of influence. This means bringing people from small "p" groups and big "P" institutions together to work on their shared "c": community.

Throughout the book, we have drawn attention to the many ways these political dynamics manifest. In the case studies, we heard from landholders that had organised to take control of their wild dog problem and in doing so, demonstrated how their collective effort had transformed their communities' understanding of a complex problem, generating potential spill-over effects for broader community development in those locations. For example, the strong social network developed at Mt Mee spilled over into individual efforts to welcome newcomers to the community.

> In the practitioner profiles, we saw a demonstrated commitment to share power through principles of collaboration and inclusiveness. The practitioners personal experience convinced them that addressing complex public issues such as pest management required a flexible engagement practice in order to increase the likelihood of community action.

These examples demonstrate that if community engagement is seen as part of a broader project of democratic capacity-building, it is possible to move beyond prescriptive best practices to foster the enabling settings for community action on a range of challenging and complex issues.

Achieving successful collective action requires collective effort. This effort needs to combine awareness of the "softer" sides of community engagement, such as relationship building and conflict resolution, with the harder stuff such as addressing structural power imbalances, institutionalised disadvantage and discrimination. For community members to come together and achieve collective action, they need the skills to deliberate, take decisions and build a collective pathway for action. In this way, community pest management should be seen as a fundamentally democratic endeavour. We suggest that incorporating democratic theory in the design and implementation of community engagement activities may deliver benefits beyond action on pest species; and that anyone can be part of this. Just as this idea has implications for policy makers, it also has implications for community members. Landholders and engagement practitioners do not need to think in terms of democratic theory to work democratically, as they can engage in the everyday politics of valuing diverse perspectives, involving all stakeholders, and listening to others.

Interaction and Intersection

Throughout this enquiry, two concepts emerged as fundamental to the lived experience of community pest management. These were the importance of *interaction* as a foundation for relationship development, trust building, community strengthening and personal support; and the ability to facilitate *intersection* among different values, objectives and interests.

> In combination, creating opportunities to increase *interaction* and encourage *intersection* made it possible for individual citizens to come together across their differences and take collective action. By *intersection* we mean that people in communities work together to find common ground. When people see *intersection* as possible, they can then find ways of working collectively.

In identifying these two key conceptual drivers, we return again to our philosophy of community engagement which draws from theories of political economy and community development. In this paradigm, the starting point for community is the individual citizen, who holds diverse values and specialised knowledge informed by their personal experience of politics, power and place. Individuals are then positioned within a web of social interactions, which create and reinforce the interdependencies that characterise *human,* as opposed to pest species ecology (McKinney and Kemmis 2011). No individual can manage a pest species alone. These interdependencies are the driving motivation for individuals to come together, even when they hold different views or values. In the narratives, we saw instances of community interdependence playing out in the pest management context, leading to interaction between diverse perspectives, and requiring an intersection of values and beliefs in order to shape a possible pathway to collective action.

As discussed throughout the book, community pest management requires trusting interpersonal relationships between those individuals who need to work together to achieve their objectives. The *interactional* element of building trusting relationships were illustrated in the practitioner profiles in Part I, as individuals made a commitment to one-on-one visits, late-night phone calls, being receptive and available, but also proactive, asking the right questions, and being interested in others' lives. The *intersectional* element was illustrated by commitments to working collaboratively to develop shared agendas, objectives and actions in order to build a shared purpose and enhance trust amongst community members and other stakeholders. The use of participatory planning processes, such as the nil-tenure approach, are helpful tools for facilitating both interaction and intersection, as they bring people together to share information and reach a common understanding of the challenge they face. These strategies were highlighted in the wild dog case studies in Part II.

Bringing Theory to Practice
At the start of the book we were transparent about our intention to strengthen community engagement practice without developing "best practice" guidelines or how-to instruction manuals.

> For us, best practice in community engagement emerges from intuitive improvisation grounded in theory.

While there are some overarching factors that guide human behaviour and practice, specific behaviours and practices vary across time, space, and place. Rather than attempt to prescribe best practices, we prefer to highlight the theories, concepts and principles that enable practitioners to develop their own engagement practice. Our aim is to equip practitioners with the confidence to address this variation through a strong personal and professional understanding of their work with communities. This requires an integration of theoretical knowledge with practical experience, in order to move beyond technocratic "best practices" towards a highly effective and fundamentally transformative practice of community engagement.

The stories captured in this collection illustrate our argument that attempts to achieve community pest management can be strengthened by this integration of practice and theory. Constructed from the transcripts of narrative interviews, the 12 practitioner profiles in Part I of this book and the three case studies in Part II contain a wealth of narrative perspectives on community pest management. In each section, we made some observations about the practices and theoretical implications that emerged from our reading of these narrative accounts. In Chap. 16 we characterised the practitioners as change agents, engaged in both personal and professional efforts to develop a practice of community engagement that would share power and build capacity towards a collective community effort. In Chap. 21 we considered the interaction between community dynamics and decision-making structures in wild dog management groups, observing the link between community context and willingness to work collectively. Throughout these narrative accounts, we have listened for instances where individuals' personal philosophies, experiences and biases have connected with theoretical insights from community development, political economy and sociology. These insights include:

- recognition of the value inherent in diverse values and the capacity to work across difference;
- awareness of power differentials and the need for active management of these dynamics in any community engagement efforts;
- managing the tension between expertise and local knowledge in understanding and responding to threats; and

- embodying the knowledge that social and environmental change begins with individual reflection and learning.

We believe that this interaction between theory and practice is fundamental to effectiveness, because it equips us with the knowledge required to situate our community work in a broader context of social change and community development. It enables us to reflect on our experience and seek new knowledge as required, in a process of continual learning that can sustain our efforts over time.

In Chap. 16 we also identified a suite of practical lessons that can support practitioners and community leaders in their work. These lessons include:

- the importance of listening to build rapport and a nuanced understanding of the issue facing the affected community;
- having a clear drive to make a difference to social and environmental outcomes;
- understanding the political dimensions of community engagement and how these are reflected in practitioner choices;
- the value of building common ground through sharing experiences with affected communities;
- the personal and professional benefits that come from a willingness to critically reflect and learn from experiences (both positive and negative); and
- the wisdom to strive for a sustainable balance between personal effort and community ownership.

The wild dog management group case studies illustrated many of the practice lessons and dilemmas raised in the practitioner profiles. For example, the "struggle for change" that characterised the practitioner profiles was the backdrop against which the case studies' efforts to develop a collective response to the wild dog threat played out, as community circumstances changed and new strategies were required.

This ability to recognise the need for change and the capacity to work positively to effect it was a common narrative thread. It supports our interpretation that achieving the aspiration of "shared responsibility" through community pest management is a political and social change project that requires profound political shifts to move power to, and share it with, the affected community. While a seemingly simple notion, this shift can be troubling for many government officials, scientific experts and community members alike.

As we noted in Chap. 1, community pest management is underpinned by the interaction between public and private good interests. Determining where the responsibility lies for managing established pests requires a careful balancing of these interests. Achieving this balance requires a theory and practice of community engagement that pivots on the ability to facilitate respectful and productive interaction, which then enables the intersection of diverse values. These skills are essential when one is faced with difficult decisions in complex contexts. A philosophical commitment to democratic practice facilitates this interaction and intersection, and enables policy-makers, scientists, practitioners and community members to combine specialised knowledge with a more nuanced understanding of public sentiments, values and localised contexts. This can lead to robust and legitimate strategies for action that can inform management decisions.

Future Directions

As we approach the end of this narrative enquiry, we consider some future directions for community pest management research.

- **Listen to diverse voices**: As we noted in Chap. 3, the practitioner sample had some demographic gaps, with practitioners all located on the eastern seaboard of Australia, only two women in the collection and no indigenous land managers. This represents an opportunity for future studies to capture these voices in this or other areas of environmental management, and we encourage readers to consider where a narrative enquiry might be useful for accessing these missing voices. This goal of listening to people from diverse backgrounds and learning from the stories in other communities also extends beyond research studies to the everyday on ground management working with people.

- **Develop and appreciate practitioner skills in narrative enquiry**: Related to this is the ongoing need for work that enriches our understanding of the policy and practice settings that support individuals to organise as a community in order to take collective action. The nuances that emerge through narrative enquiry offer important insights into the dynamics that might constrain or undermine community efforts to come together and take collective action. This work can also help develop critical understanding of the features of collective action and the distinctions between working towards this objective and other forms of community engagement. In-depth analyses are useful for linking particular management strategies and engagement approaches to specific contexts, and they can be enhanced by a more systematic approach to data collection, enabling broader conclusions about the settings that support and sustain collective action (Poteete and Ostrom 2008). We recommend practitioners and researchers develop their skills in soliciting and analysing community narratives in order to develop a practical knowledge of collective action and most importantly, share these stories and insights with others who are working to achieve the same objective.

- **Experiment with community development as a model for pest management**: Learning from practitioners can pave the way for new policy and funding models that acknowledge the key role of community engagement and provide adequate time and resources to enable this undertaking. We suggest that reframing this work as "community development" would enable practitioners and community members to position their efforts within the broader context of their contribution to the overall health of their community. Community development is associated with the use of participatory techniques for planning and evaluation, and may support the devolution of power from government to community members through a range of different mechanisms such as partnerships, collaborations or co-management arrangements. This makes it a political act, which will encounter resistance—however we would like to see more social research that articulates these political dimensions in order to reinvigorate the "narrative" of community action. All stakeholders can play a role in this reframing through their actions and their conversations.

- **Integrate narrative insights into traditional biophysical and technocratic research design**: This work has deepened our appreciation of narratives and narrative inquiry as a research approach. We believe this approach is central to understanding individual and collective human behaviour, and provides a useful complement to deductive and quantitative analyses. We believe it affords insight into the variance in human behaviour in particular contexts and thus has central relevance to public policy and program design at both local and larger levels. Using stories to communicate enables people to bridge differences in values, beliefs, ideology, ways of knowing, approaches to problem-solving and the myriad other factors that divide human beings and restrict their positive interaction. Flyvbjerg (2001, p. 3) points out that social science enables "reflexive analysis and discussion of values and interests, which is the prerequisite for an enlightened political, economic, and cultural development in society". We encourage greater experimentation with narrative enquiry research strategies, including storytelling, as vehicles for better understanding complex societal problems, including who is affected by initiatives for addressing those problems, how people are affected, and what knowledge, experiences and perspectives people they can bring to carrying out or rethinking existing initiatives.

An Invitation to Learn

In Chap. 3, we urged the reader to accept the invitation issued by practitioner Darren Marshall to join him on a learning journey. Throughout the book we have continued to reissue this invitation to the reader because the purpose of the collection is not only to articulate a community engagement practice, but also to strengthen this practice through stimulating critical reflection. We encourage you to revisit our analysis of the practitioner profiles and the wild dog case studies, consider how our "take" agrees or diverges from your own understanding, and how this comparison might inform you and your practice.

This book has been written for individuals who work to achieve collective action. Our audience includes policy-makers, program designers, wild dog controllers, landholders and all other stakeholders who have an influential role to play in the "guiding coalition" that supports community pest management. We believe all stakeholders involved in community engagement and collective community action of any kind, can learn from these stories of the realities of designing, funding, supporting and implementing community pest management. We hope these narratives will pave the way for new policy and funding models that acknowledge the central role of community members and community engagement practice in driving social and political change over time. A key message from these stories is that listening, thinking and working with community are time consuming activities that require adequate support over extended timeframes if they are to be effective and the desired outcome is to be achieved.

In conclusion, we would like to reaffirm our conviction that for community pest management to succeed, both social and biophysical knowledge is required, because neither alone is enough. This book has shown that it is possible to connect these bodies of knowledge through the lived experience of community pest management. In presenting these narratives, we have looked for models of interaction and intersection that support our understanding of community pest management as dynamic, evolving and essentially political. These practice stories contribute valuable knowledge about the social and political dynamics that must be addressed if the vision of "shared responsibility" for pest management is to be achieved. This returns us to our starting point, and we take this final chance to reiterate that:

- pest management is a community problem, that requires collective action in order to achieve best results across the landscape; and that
- collective action requires people to work together to develop a shared vision and commitment, to the problem and to each other, in order for that action to be sustained over time, in response to the persistent nature of pest species.

Working to achieve community pest management is essentially a political act. We encourage all individuals with a stake in community pest management to consider the political dimensions of their efforts to exchange knowledge, broker agreement and develop on ground action plans. This may require relinquishing personal or organisational power, addressing conflict and disagreement, or modelling an engagement practice that supports critical reflection and humility. For us, the capacity to engage in meaningful dialogue across our differences is the foundation for collective community action. In this book, we have attempted to model this approach through the regular invitation to the reader to make their own analysis of the text. We hope this invitation stimulates an ongoing dialogue that will make the most of the rich resources available in these collected narratives.

References

Boyte, H. C. (2004). *Everyday politics: Reconnecting citizens and public life*. Philadelphia, USA: University of Pennsylvania Press.

Diekman, B. (Ed.). (1994). *Unwanted aliens: Australia's introduced animals: Proceedings of a seminar held at the Australian museum, Sydney, 23–24 september 1994*. Sydney, NSW: Nature Conservation Council of NSW.

Flyvbjerg, B. (2001). *Making social science matter: Why social inquiry fails and how it can succeed again*. Cambridge: Cambridge University Press.

Mayer, F. (2014). *Narrative politics: Stories and collective action*. New York: Oxford University Press.

McKinney, M., & Kemmis, D. (2011). Collaboration and the ecology of democracy. *Human Dimensions of Wildlife, 16*(4), 273–285.

Poteete, A. R., & Ostrom, E. (2008). Fifteen years of empirical research on collective action in natural resource management: Struggling to build large-N databases based on qualitative research. *World Development, 36*(1), 176–195.

Glossary

There are a number of Australian slang terms that appear in the narratives. Brief definitions are provided here for our international readers.

Bugger Exclamation of frustration
Bush telegraph The informal lines of communication in a rural community
Cocky (pl. cockies) An Australian farmer, usually small scale
Cop it Endure, put up with
Dodgy Unstable or unreliable
Dogger A technician with expertise in dog control such as trap setting and dog baiting
Flack Complaints or criticism
Get stuffed Instruction to go away, often said in outrage or exasperation
House paddock Fenced area closest to the house
Ripping The practice of destroying rabbit warrens
Show Local agricultural event
Swag A canvas bedroll used for camping
Ute A utility vehicle with a tray on the back
Whinging Complain

Other definitions:

1080 A metabolic poison commonly used in dog control baits
Barrier fence A fence designed to keep particular species out
BESTWOOL/BESTLAMB A sheep management program operated by industry body Australian Wool Innovation
Landcare A national program focused on achieving landscape restoration and regenerative agriculture through local landholder groups
Nil-tenure Nil-tenure is a planning approach that takes a holistic view of the landscape, unimpeded by land ownership conventions. Landholders are encouraged to think about common threats as not constrained by property boundaries and to take action collectively in response to the shared problem

© Springer Nature Singapore Pte Ltd. 2019
T. M. Howard et al., *Community Pest Management in Practice*,
https://doi.org/10.1007/978-981-13-2742-1